Microwave Non-Destructive Testing and Evaluation

NON-DESTRUCTIVE EVALUATION SERIES

Volume 4

Non-destructive evaluation now has a central place in modern technology both as a means of evaluating materials and products as they are manufactured and for confirmation of fitness for purpose while they are in use.

This series provides in-depth coverage of the wide range of techniques that are now available for the non-destructive evaluation of materials. Each volume will contain material that is relevant to final year undergraduates in Engineering, Materials Science and Physics in addition to post graduate students, experienced research workers, and practising engineers. In some cases they will be written with taught courses in mind, while other texts will be for the qualified engineer or scientist who wishes to become familiar with a new topic at research level.

Series editor

Professor W. Lord
Department of Electrical and Computer Engineering
Iowa State University
Iowa
USA

Volume 1: Halmshaw: Industrial Radiology: Theory and Practice
ISBN: 0-412-62780-9

Volume 2: Blitz/Simpson: Ultrasonic Methods of Non-destructive Testing
ISBN: 0-412-60470-1

Volume 3: Blitz: Electrical and Magnetic Methods of Non-destructive Testing
ISBN: 0-412-79150-1

Microwave Non-Destructive Testing and Evaluation

by

REZA ZOUGHI

Applied Microwave Nondestructive Testing Laboratory (*amntl*)
Electrical and Computer Engineering Department
Colorado State University
Ft. Collins, Colorado
U.S.A.

KLUWER ACADEMIC PUBLISHERS
DORDRECHT / BOSTON / LONDON

A C.I.P. Catalogue record for this book is available from the Library of Congress.

ISBN 0412-625008

Published by Kluwer Academic Publishers,
P.O. Box 17, 3300 AA Dordrecht, The Netherlands.

Sold and distributed in North, Central and South America
by Kluwer Academic Publishers,
101 Philip Drive, Norwell, MA 02061, U.S.A.

In all other countries, sold and distributed
by Kluwer Academic Publishers,
P.O. Box 322, 3300 AH Dordrecht, The Netherlands.

Printed on acid-free paper

Printed in the Netherlands.

To *Molouk* and *Valiollah*

Contents

Foreword

Microwave and millimeter-wave non-destructive testing and evaluation (NDT&E) is generally understood to mean using high-frequency electromagnetic energy to inspect and characterize materials and structures. In spite of possessing some distinct advantages in certain applications compared to other NDT&E techniques, microwave NDT&E has only found limited practical application during the past 45 years. These advantages include lack of a need for contact between the sensor and the object being inspected, the ability to penetrate dielectric materials, and superior sensitivity to certain material constituents and flaws.

One factor contributing to this minimal acceptance by the NDT&E community has been a generally poor understanding in this community of the theory and practice that underlie the technology. This situation exists partly because of a paucity of microwave NDT&E textbook and reference material. Some chapters, reviews, and books aimed at filling this need have been published in the past but, for the most part, this material is based on the use of older microwave technology. However, during the past ten years great strides have been made in terms of the cost, size, and ease of use of microwave components. In addition, recent advances in modeling and measurement techniques have expanded the range of applications for microwave NDT&E. Such applications include inspecting modern materials such as composites, detecting and characterizing surface flaws, and evaluating the compressive strength of cement structures. These advances have created an urgent need for up-to-date textbook material on this subject.

Professor Zoughi, his colleagues, and his students have been at the forefront of these recent advances in microwave NDT&E. Therefore, I believe that this textbook will become a primary reference in the field and it should be exceedingly useful to those who wish to learn more about current theory and practice in microwave NDT&E. I strongly recommend it to both students and practitioners.

Alfred J. Bahr, Ph.D.
Principal Scientist
SRI International
Menlo Park, California
USA

Colour Plates

Fig. 5.8 A phase scan intensity image of the composite panel shown in Fig. 5.6 at a frequency of 10.5 GHz and when in contact (Qaddoumi *et al.*, 1995a).

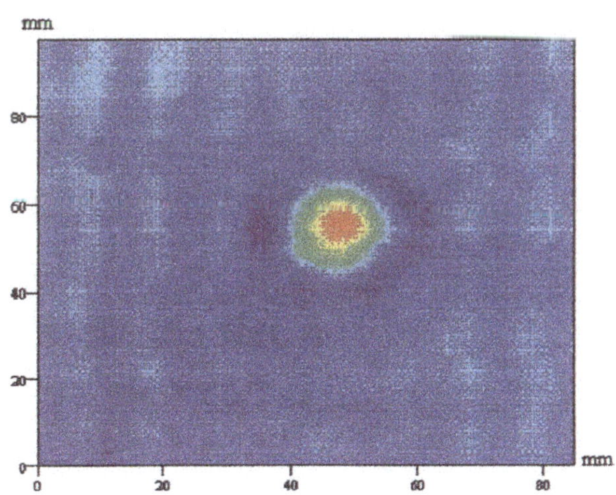

Fig. 5.9 A phase scan image of the composite panel shown in Fig. 5.6 at a frequency of 10.5 GHz and when in contact (Qaddoumi *et al.*, 1995a).

Fig. 5.10 A phase scan intensity image of the composite panel shown in Fig. 5.6 at a frequency of 10.5 GHz and at a standoff distance of 9 mm (Qaddoumi *et al.*, 1995a).

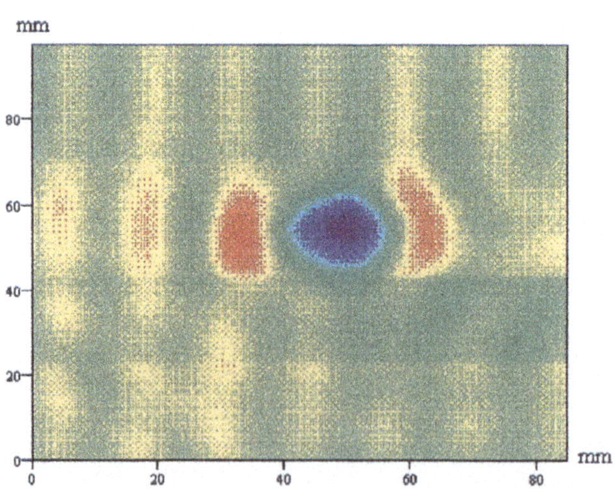

Fig. 5.11 A phase scan image of the composite panel shown in Fig. 5.6 at a frequency of 10.5 GHz and at a standoff distance of 9 mm (Qaddoumi *et al.*, 1995a).

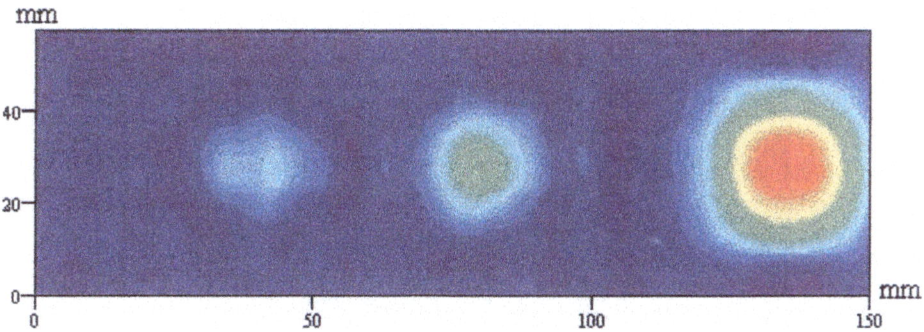

Fig. 5.12 An amplitude scan image of a thick glass reinforced polymer composite panel with three aluminum inclusions at a frequency of 10.5 GHz and when in contact (Qaddoumi *et al.*, 1995a).

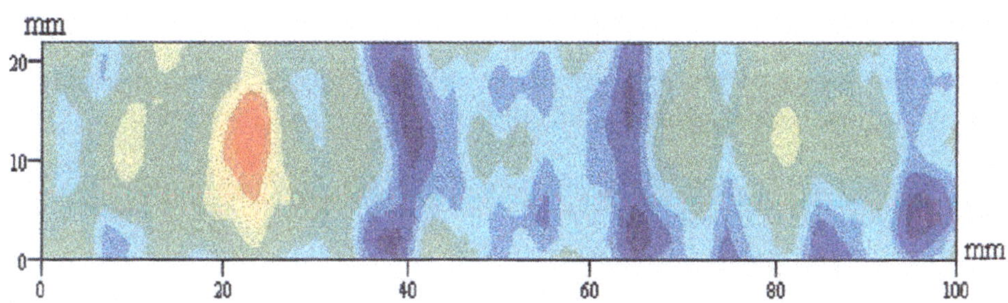

Fig. 5.14 Image of the composite panel shown in Fig. 5.13 at a frequency of 24 GHz and when in contact (Qaddoumi *et al.*, 1995a).

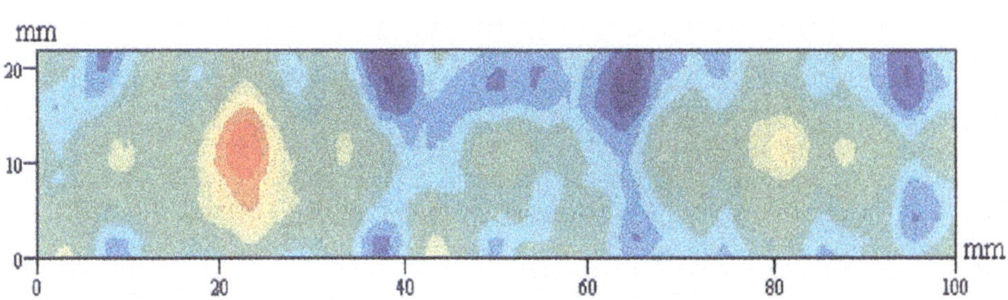

Fig. 5.15 Image of the composite panel shown in Fig. 5.13 at a frequency of 24 GHz and at a standoff distance of 0.5 mm (Qaddoumi *et al.*, 1995a).

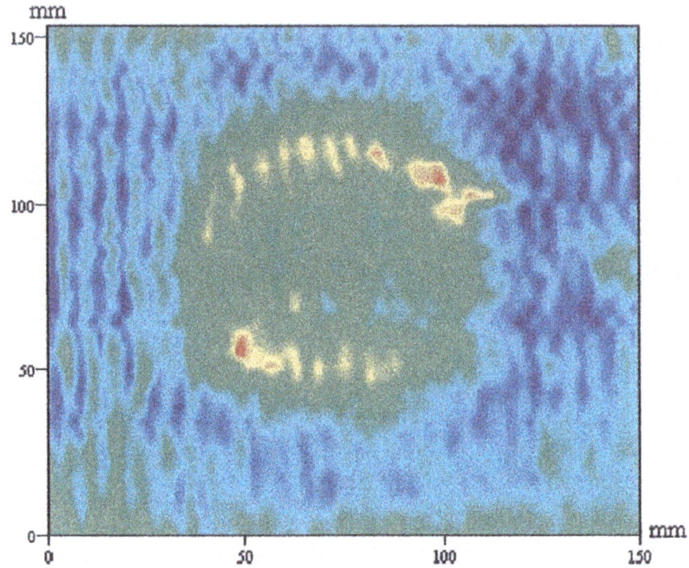

Fig. 5.16 Image of the disbond simulated by a circular mylar sheet in a sandwich composite (Ganchev *et al.*, 1995b).

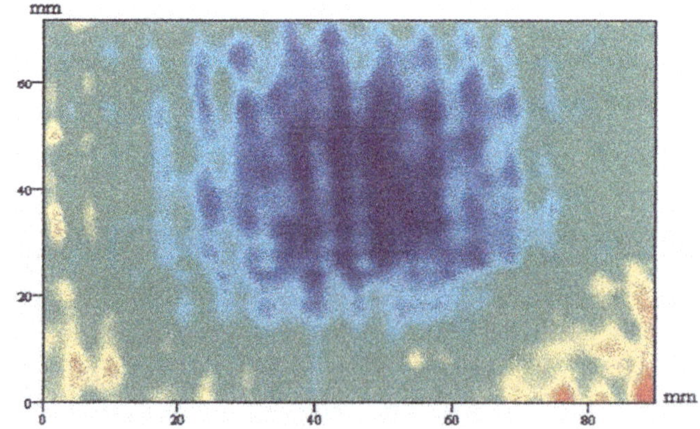

Fig. 5.17 Image of the impact damage in a sandwich composite (Ganchev *et al.*, 1995b).

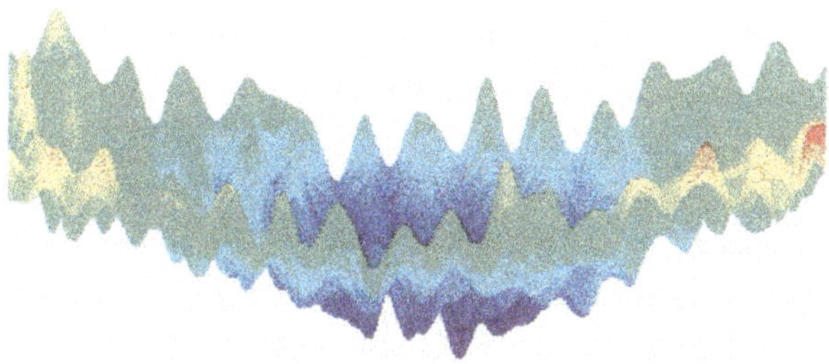

Fig. 5.18 Intensity plot (3-D surface plot) of the impact damage in a sandwich composite (Ganchev *et al.*, 1995b).

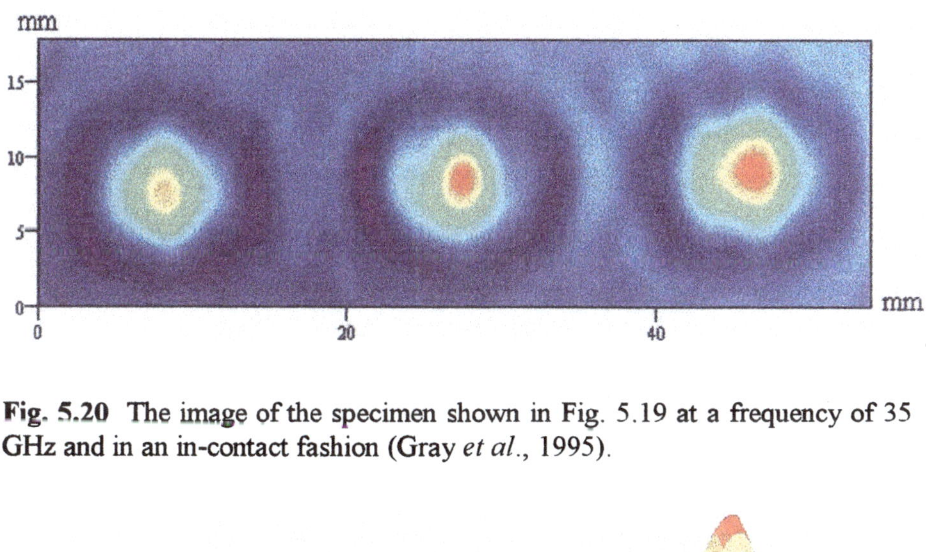

Fig. 5.20 The image of the specimen shown in Fig. 5.19 at a frequency of 35 GHz and in an in-contact fashion (Gray *et al.*, 1995).

Fig. 5.21 Intensity plot of the image shown in Fig. 5.20.

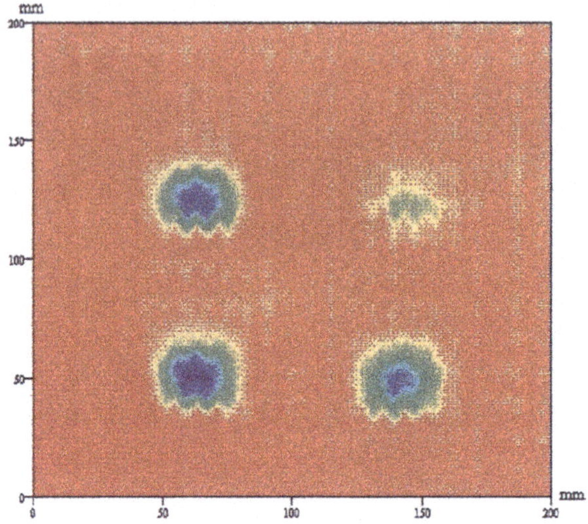

Fig. 5.23 Image of the fiberglass specimen in Fig. 5.22 at a standoff distance of 4 mm and a frequency of 24 GHz (Qaddoumi, Ganchev and Zoughi, 1996).

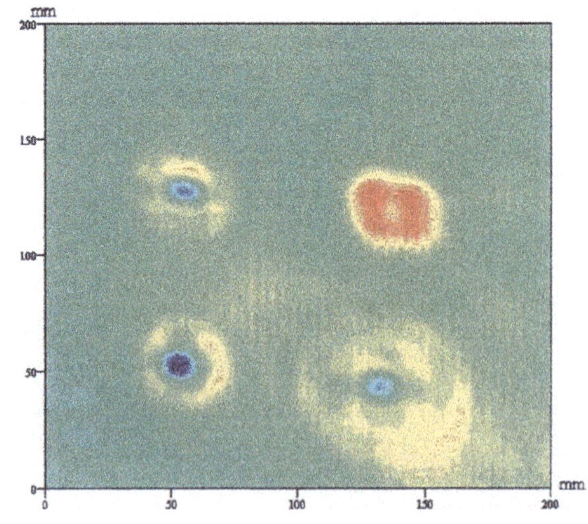

Fig. 5.24 Image of the fiberglass specimen in Fig. 5.22 at a standoff distance of 10 mm and a frequency of 10 GHz (Qaddoumi, Ganchev and Zoughi, 1996).

Fig. 5.25 A 40 mm by 40 mm area of rust on a steel plate (Qaddoumi, Shroyer and Zoughi, 1997).

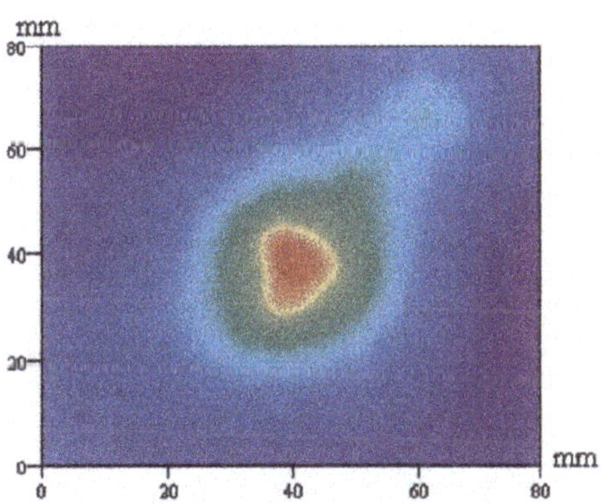

Fig. 5.26 Image of the rust shown in Fig. 5.25 under 0.292 mm of paint, at a frequency of 24 GHz and a standoff distance of 4 mm (Qaddoumi, Shroyer and Zoughi, 1997).

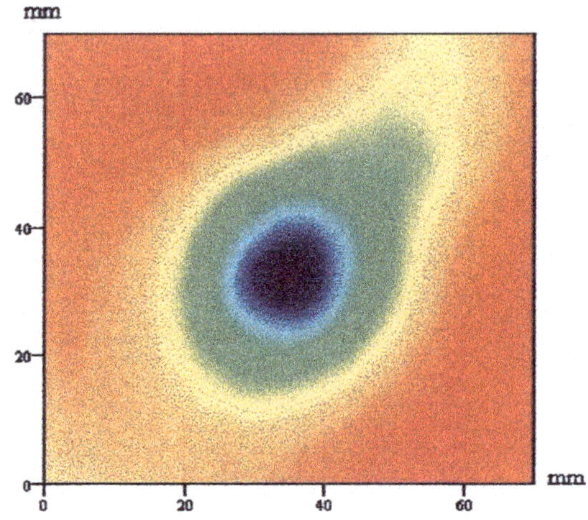

Fig. 5.27 Image of the rust shown in Fig. 5.25 under 0.267 mm of paint, at a frequency of 24 GHz and a standoff distance of 12.5 mm (Qaddoumi, Shroyer and Zoughi, 1997).

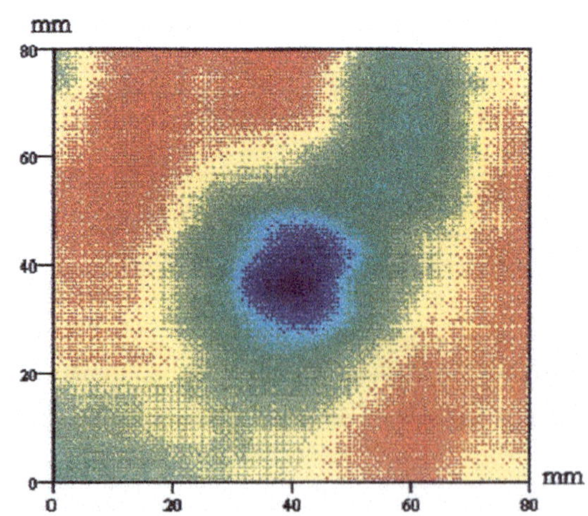

Fig. 5.28 Image of the rust shown in Fig. 5.25 under 0.292 mm of paint, at a frequency of 10 GHz and a standoff distance of 2 mm (Qaddoumi, Shroyer and Zoughi, 1997).

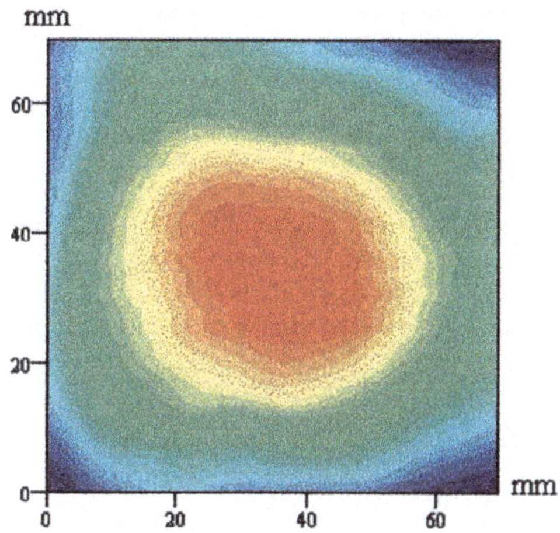

Fig. 5.29 Image of the rust under 25.4 mm of a low-permittivity synthetic rubber composite, at a frequency of 10 GHz and a standoff distance of 2 mm.

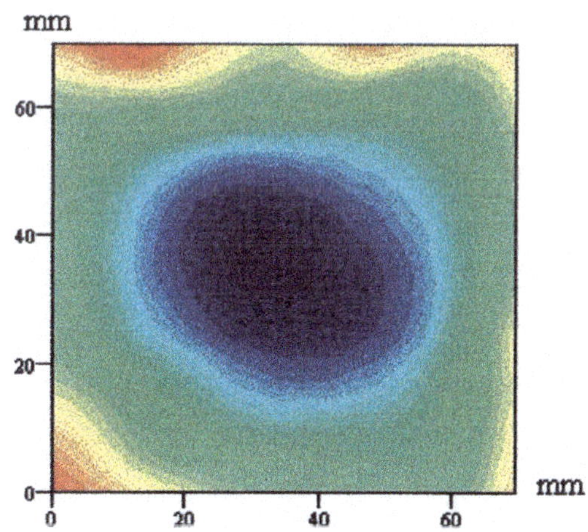

Fig. 5.30 Image of the rust under 25.4 mm of a low-permittivity synthetic rubber composite, at a frequency of 24 GHz and a standoff distance of 6 mm.

CHAPTER 1

Introduction

1.1 GENERAL REMARKS[1]

Microwave non-destructive inspection techniques have a long history dating from the early 1950s, with a strong flurry of activities in the 1960s and 1990s (Bahr, 1982). However, these techniques are still not widely known in the non-destructive testing (NDT) community and are often referred to as 'emerging techniques' or 'others'. It is only during the past few years that some NDT conferences have allocated a whole session or two solely to the topic of microwave NDT. For those involved in the research and development of microwave and millimeter wave NDT techniques who have long appreciated the advantages of such techniques, this increased visibility is a welcome and encouraging change.

It appears that unlike the 'NDT community', the 'microwave community' is generally more aware of the applications of microwaves for non-intrusively inspecting materials and structures. However, articles published in microwave oriented journals (not NDT) do not always receive the attention of NDT practitioners. Furthermore, most of these scientific investigations may not always attack an NDT problem directly, but the approaches developed by them can be modified and expanded to include practical NDT problems. The theoretical development of an antenna admittance variations in layered plasma is a good example of this. Although most of such problems have dealt with understanding the characteristics of an antenna on a spacecraft during re-entry into the atmosphere, a layered plasma is very similar to a layered dielectric composite. There are also new microwave and millimeter wave theoretical and experimental developments specifically for NDT purposes that have taken place in the last few years (in addition to research in the 1970s and

[1] Portions of this section are reprinted with permission from *Materials Evaluation*, vol. 53, no. 4, pp. 461-62, Copyright 1995 © The American Society for Nondestructive Testing, Inc.

early 1980s). The inspection of new dielectric materials such as plastics, ceramics and other composites represents an example of an important new potential application area for microwave and millimeter wave NDT.

The ever expanding materials technology by which lighter, stiffer, stronger and more durable electrically insulating composites are replacing metals in many applications demands alternative inspection techniques to the existing NDT approaches. This is due to the fact that the existing standard and well established NDT techniques (developed primarily for inspecting metallic structures) may not always be capable of inspecting these composites.

Microwave signals penetrate inside dielectric (electrically insulating) media easily. The limitation on the depth, from which useful information may be obtained, is primarily a function of the loss factor of the dielectric material (ability to absorb microwave energy), the frequency of operation and the receiver sensitivity (noise floor). Microwave measurements can be conducted in-contact or non-contact while operating on one side of a material or using its both sides (reflection or transmission techniques). Microwave NDT techniques are sensitive to geometrical and dimensional properties of a medium or a defect. Polarization properties of microwave signals can be used to increase measurement sensitivity to defects of a certain orientation. The evaluation of the properties and composition of mixtures including the effect of curing in chemically produced composites is also possible. It is not necessarily true that because microwave signals have wavelengths in the centimeter range, the resolution obtained using these signals cannot be better than a large fraction of the operating wavelength. For instance, in monitoring thickness variation in dielectric slabs and coatings, if resolution is considered to be the smallest thickness variation that is possible to be detected, then near-field microwave NDT techniques have shown measurement resolutions of a few micrometers at 10 GHz (wavelength of 3 cm in free-space). If resolution, in a given application, is considered to be the smallest spatial distance between two defects so that they can be individually detected, then near-field microwave and millimeter wave techniques have provided resolutions of better than a tenth of a wavelength and in some recent publications on microwave microscopy resolutions of better than one fiftieth of the operating wavelength have been achieved. This is due to the fact that lateral resolution in the near-field of a probe such as an open-ended waveguide aperture or an open-ended coaxial line is dictated by the probe dimensions rather than the operating wavelength. It has also been shown that an increase in the operating frequency does not necessarily render higher measurement resolution in all cases. Also, at around 10 GHz fatigue cracks on metal surfaces with widths of less than ten microns have been detected. In addition, spatial resolution is a function of the microwave sensor used (e.g. open-ended waveguides, open-ended coaxial lines, cavity resonators, etc.). Statements made in conjunction with radar resolutions do not always apply to microwave NDT techniques since most of NDT applications are conducted in the near-field of a probe as opposed to its far-field. Furthermore, changes in the reflection coefficient properties for a given microwave sensor and defect are very different in the near-field compared to the far-field.

Hardware systems for microwave NDT applications need not be expensive. It is true that if laboratory test equipment is used to conduct measurements, the cost will be high. However, hardware for a specific application can be developed and built to be relatively inexpensive, simple in design, hand-held, battery operated, operator friendly and operate on an on-line basis. In a majority of microwave NDT applications where detection is the primary objective, there is very little to no need for complicated post signal processing. Capability of providing real-time information makes these techniques suitable for on-line industrial applications. In most applications, the operator need not be a microwave expert to conduct microwave NDT measurements once a system has become operational. Since, the required operating power for most NDT applications (excluding ground probing radars and microwave heating sources) are in the few milliwatts range, and the majority of these techniques are conducted in near-field with a high degree of frequency selectivity (narrow band CW measurements), they do not cause any EMI and are not affected by EMI.

Large scan areas can be accommodated with an array of sensors. Microwave signals do not penetrate inside conductors and graphite composites. However, this is not always a limitation since surface features such as surface cracks and defects in metals and impact damage on graphite composites can be inspected along with evaluating the properties of dielectric coatings (thickness and material composition) on top of metals and graphite composites. One of the most attractive aspects of microwave NDT techniques is the availability of many different probes/sensors. Some of these may render better results than others for a specific application. In addition, optimization of system parameters for obtaining more sensitive results makes microwave NDT techniques very powerful.

Once the underlying theoretical foundation of the interaction of microwaves with a given medium is understood and modeled, one may develop electromagnetic codes to predict the outcome of a measurement in order to optimize it for obtaining the highest possible measurement sensitivity. It is argued that unless one has a reasonably sound understanding of the fundamental steps involved in the interaction of microwaves with various media and a thorough understanding of how to manipulate various microwave sensors for measurements purposes, the so-called tinkering with microwave sensors may not render desired and acceptable results.

Areas that may benefit (some already have) from using microwaves NDT techniques are:

1. Composite inspection:
   - accurate thickness measurement of coatings, single dielectric slabs, layered dielectric composites made of plastics, ceramics, wood and any other type of dielectric materials,
   - minute thickness variation detection of each layer in these dielectric media,
   - disbond, delamination and void detection and thickness evaluation in stratified and half-space of dielectric composites,
   - potential for depth determination of a disbond in a stratified composite structure or delamination in a thick dielectric material,
   - detection and evaluation of rust and corrosion under paint and thick stratified laminate composite coatings,
   - inspection of thick plastics and glass reinforced composites for detection of localized and area-extensive interior flaws,
   - detection and evaluation of fiber bundle orientation or mis-orientation and breakage in reinforced composites during production (process control) and while in service,
   - impact damage detection and evaluation in reinforced composite structures including graphite composites.

2. Dielectric material characterization:
   - accurate dielectric material property characterization,
   - accurate constituent characterization, such as constituent volume content determination and constituent dielectric property evaluation in dielectric mixtures,
   - detection and estimation of porosity in ceramics, thermal barrier coatings, polymers, glass, rubber, etc.,
   - detection and evaluation of curing in chemically reactive materials including resin binders, epoxy, cement based materials, rubber, etc.,
   - correlating electrical properties of materials, such as dielectric characteristics, to their physical and mechanical properties, such as compressive strength of concrete, detection of stress and strain in glass fibers, etc.

3. Metal surface inspection:
   - detection of stress-induced fatigue (surface) cracks in metals,
   - potential for crack sizing (all three dimensions),
   - crack tip location determination,

- detection of cracks masked by various dielectric coatings such as paint and rust as well as detection of cracks filled with rust, dirt, paint, salts and other dielectric fillers. Furthermore, cracks under coatings may be detected easier than exposed cracks. Hence, no need for paint removal, particularly when its disposal poses environmental concerns (such as lead-based paint),
- optimization capability for obtaining increased measurement sensitivity,
- metal surface roughness evaluation,
- accurate metal surface profiling,
- surface flaw detection and evaluation in rolled metal sheets such as rolled steel and aluminum products.

4. Microwave imaging:
   - image production of localized and distributed interior and surface flaws,
   - near-field as well as far-field (uniform plane-wave) imaging techniques, employing various image reconstruction techniques, and in some cases achieving near real-time microwave images.

5. Medical and industrial applications:
   - detection of unhealthy skin patches,
   - near real-time imaging of limbs,
   - hyperthermia,
   - moisture and humidity detection in timber, wood, textiles, grains, foods, etc.,
   - timber excess moisture and knot detection,
   - resin property evaluation in textile production,
   - sintering in production of ceramics,
   - hardened cement paste water-to-cement ratio evaluation and correlating near-field microwave reflection properties of cement paste to its compressive strength,
   - detection of porosity in mortar and evaluating mortar water-to-cement as well as sand-to-cement ratios,
   - concrete inspection for evaluating its constituents, detecting and locating reinforcing steel bars, deterioration evaluation, constituent segregation detection, aggregate volume content and distribution determination, chloride detection, correlating near-field microwave reflection properties of concrete to its compressive strength, etc.,
   - evaluating concrete water-to-cement ratio and its cure state monitoring,
   - detection of grout in masonry structures,

- radars, including ground penetrating radar (GPR) for detecting buried objects,
- radiometry (a very sensitive microwave receiver tuned to the natural radiation from a material at a particular frequency),
- and finally, new applications are continuously being discovered.

The time is right for renewed vigor in applying microwave NDT techniques to new inspection problems. A fresh sustained effort in this area should result in many new findings. Many investigators believe that, given the same level of effort enjoyed by other NDT techniques during their development, microwave NDT will find a prominent place in a wide realm of applications. In some cases it will prove to be the technique of choice. In other cases it will prove to be a useful complement when used in conjunction with other methods. It is the author's opinion that microwave sensor fusion with other NDT techniques as well as an increased industrial utilization of microwave based NDT methods will be the two areas that will significantly impact the future of microwave and millimeter wave NDT.

There are several other issues associated with the use and application of microwave NDT techniques that must also be mentioned. These issues include, but are not limited to, technical cultural differences and lack of familiarity with these techniques by NDT practitioners (accustomed to using 'standard NDT techniques'). At times, having unreasonable expectations when using these techniques, specially if microwave NDT techniques are deemed the only solution to a given problem. The need for understanding that many of these techniques are optimally designed to provide solution for a certain application, and for use in other applications (sometimes even for similar applications) they may need to be re-designed/re-optimized (Bolomey, 1996).

Throughout this text the term 'NDT' will be used when testing for defects or state of a material. However, when interested in evaluating the properties of a material the term 'NDE' signifying non-destructive evaluation will be used.

1.2 MICROWAVE FREQUENCY BANDS

During the past forty years, various letter designations have been assigned to various microwave and millimeter wave frequency bands. These designations have been assigned either by the military or by technical and standard setting institutions. Table 1.1 lists the commonly used microwave and millimeter wave band letter designations and their respective operating frequency ranges and the waveguide designation for each band.

Table 1.1 Microwave and millimeter wave frequency band letter designations, the respective operating frequency of rectangular waveguides and their designation.

Letter designation	*Frequency range (GHz)*[1]	*Waveguide designation*[2]
L	1.12-1.70	WR-650
R	1.70-2.60	WR-430
S	2.60-3.95	WR-248
G	3.95-5.85	WR-187
J	5.85-8.20	WR-137
H	7.05-10.0	WR-112
X	8.20-12.4	WR-90
Ku	12.4-18.0	WR-62
K	18.0-26.5	WR-42
Ka[3]	26.5-40.0	WR-28
Q	33.0-50.5	WR-22
U	40-60	WR-19
V	50-75	WR-15
E	60-90	WR-12
W	75-110	WR-10
F	90-140	WR-08
D	110-170	WR-06
G[4]	140-220	WR-05
--	170-260	WR-04
--	220-325	WR-03
--	325-400	WR-2.8
--	400-500	WR-2.2

1 Denotes the operating frequency range within which dispersion is considered acceptable without interference from the next higher-order mode.

2 The number is related to the broad dimension of the rectangular waveguide (in 10 mils).

3 The start of the millimeter wave bands.

4 Microwave G-band (WR-187) is also referred to as H-band to avoid the obvious problem.

1.3 REQUIRED PRIOR KNOWLEDGE

This text is prepared while having two groups of interested readers in mind; namely, a) those who will use it as a graduate text solely or in conjunction with other texts, and b) those who will use it as a reference text. For the former group, this text is organized in such a way that it may be readily used in a graduate course dealing with the specific topic of microwave and millimeter wave NDT similar to the one offered at the electrical and computer engineering department at Colorado State University. Graduate students taking such a course and using this text should have a sound and fundamental knowledge of time varying electromagnetic fields and waves, transmission line theory, at least one senior or graduate level course in microwave and millimeter wave engineering including component, device and system design and integration, and finally a sound and working knowledge of graduate level analytical and numerical mathematical techniques. A basic knowledge of antenna theory and design will also be quite helpful. The author strongly recommends that instructors using this text initially present the students with a succinct yet thorough introduction to several prominent NDT methods such as ultrasonics, eddy current, magnetic particle testing, thermography, etc. for comparison purposes. For the latter group, a basic knowledge of time varying electromagnetic fields and waves would be very beneficial. This knowledge coupled with a comprehensive familiarity with NDT terminology should be sufficient to evaluate and utilize the benefits of microwave and millimeter wave NDT techniques.

1.4 ORGANIZATION OF CHAPTERS

The materials presented in this text are primarily a compilation of various research outcomes conducted during the past few years by researchers and students at the Applied Microwave Nondestructive Testing Laboratory (*amntl*) in the electrical and computer engineering department at Colorado State University. The realm of research activities in this laboratory using microwave and millimeter wave non-destructive testing techniques has been quite broad covering areas such as material characterization, layered composite inspection, metal surface inspection for cracks and surface anomalies, inspection of concrete materials, and so on.

Chapter 2 gives the basic physics behind defining an important parameter; namely, the dielectric constant. It also outlines the important features associated with this parameter regarding material characterization at microwave and millimeter wave frequency. Subsequently, an example of a simple yet versatile technique for measuring the dielectric properties of solids, liquids and powders is provided. This technique is subsequently used to determine the dielectric properties of carbon loaded rubber and its constituents in a relatively broad range of frequencies. The issue of curing and cure state monitoring is also discussed as a result of the nature of which these rubber materials are produced. The important issue of measurement accuracy associated with this technique is discussed as well. Following these discussions, the same technique is used to determine the dielectric properties of resin binder as a function of its cure state at room temperature. The resin is used in fiberglass production, hence the results of the dielectric property measurements of fiberglass specimens with various resin binder levels are discussed. Porosity in polymer plastics, thermal barrier coatings, ceramics and epoxy resin is an important parameter of interest. The same microwave dielectric measurement technique is then used to measure porosity in several polymer microballoon-filed epoxy resin specimens. Rust and corrosion accumulation under paint and other composite coatings backed by a variety of metals is an important safety concern. Rust possesses diverse material and dielectric properties. To this end, the results of the measurement of the dielectric properties of several types of rust are presented. These measurements are conducted using a different than above measurement technique the basics of which are also described. Finally, the utility of empirical dielectric mixing models for predicting the dielectric properties of carbon loaded rubber and polymer microballoon-filed epoxy resin, respectively as a function of carbon black and distributed air volume fractions is discussed.

Chapter 3 gives an overview of far- and near-filed microwave techniques for inspecting layered dielectric composites backed by free-space, an infinite half-space of a dielectric and a conducting plate. This overview is followed by a detailed derivation of the reflection coefficient characteristics of an open-ended rectangular waveguide aperture radiating into a general multi-layered dielectric composite. Subsequently, the results of this model are compared to their experimental counterparts for several practically important cases such as dielectric slab thickness variation detection, disbond detection, standoff distance and frequency optimization and disbond detection in several diverse composites.

Chapter 4 introduces two new microwave approaches, using open-ended rectangular waveguide probes, for detecting and evaluating surface cracks in metals; namely, the dominant and the higher-order mode techniques. The concept of crack characteristic signal is introduced next. The detailed steps for modeling the interaction of open-ended rectangular waveguides with a crack is provided. Subsequently, the results of an extensive set of measurements exploring the influences of various crack properties as well as measurement parameters on the crack characteristic signal are presented. These results also present crack sizing techniques as well as determining crack tip location. Finally, the results of a recent study in applying these techniques to stress-induced fatigue cracks are discussed. In addition, the potential of these techniques for crack opening determination is also presented.

Chapter 5 gives a detailed discussion of near-field measurement techniques and applications. In addition, several specific measurements in which the raster scan images of embedded defects are obtained are also presented in this chapter. The goal of this chapter is to give the reader a clear understanding of the capabilities of microwave near-field measurements and imaging.

Chapter 6 gives a brief overview of several other important recent applications of near-field microwave NDT. A discussion regarding the future of these techniques is provided as well.

1.5 ACKNOWLEDGMENT

The materials presented in this text are not certainly exhaustive. As mentioned earlier, they primarily comprise of the relatively recent research conducted by many graduate and undergraduate students, post doctorate and research associate fellows at the Applied Microwave Nondestructive Testing Laboratory (*amntl*) in the electrical and computer engineering department at Colorado State University. Other investigators' work, when relevant, have also been mentioned either specifically or in the references.

There are several individuals whose support and assistance have been invaluable in advancing the science of microwave and millimeter wave NDT at *amntl*. In particular, the author would like to specifically acknowledge professor W. Lord and Dr. G.W. Carriveau.

Others whose works have appeared in this text include: Dr. S. Bakhtiari, Dr. S.I. Ganchev, Dr. C.Y. Yeh, Dr. C. Huber, Dr. H. Abiri, Dr. N. Qaddoumi, Dr. D. Radford, Dr. K. Bois, Mr. L. Handjojo, Mr. Y. Wang, Ms. Sheila Gray, Mr. E. Ranu, Ms. R. Salem, Mr. R. Runser, Mr. A. Benally, Mr. R. Mirshahi, Mr. J. Easter, Mr. V. Otashevich, Mr. D. McColskey and many other individuals. To all, the author expresses his deepest gratitude and appreciation.

Several organizations who have provided funding for some of the research results presented in this text are The Federal Highway Administration (Dr. Richard A. Livingston), The Naval Surface Warfare Center, Welding and NDE Branch (Mr. John J. DeLoach, Jr., Mr. Lawrence M. Brown and Ms. Michele Novack), National Science Foundation (Dr. Ken Chong), Electric Power Research Institute (DR. John Maulbetsch) and several industrial organizations which have been acknowledged in papers published as a result of their support.

The author also thanks organizations, such as the Institute of Electrical and Electronics Engineers (IEEE), Springer-Verlag, and the American Society for Nondestructive Testing (ASNT), who have granted permission to reproduce copyrighted materials that have appeared in their various journals. Throughout the text wherever such copyrighted materials are used, acknowledgment has been duly provided.

Finally, it is the author's great privilege to have had Dr. Alfred J. Bahr provide the Foreword to this text.

REFERENCES

Bahr, A.J. (1982) Microwave nondestructive testing methods, Gordon & Breach, Newark, NJ.

Bolomey, J.C., (1996) Some aspects related to the transfer of microwave sensing technology. *Materials Research Society Symposium Proceedings*, 430, pp. 53-58.

Zoughi, R., (1995) Microwave and millimeter wave nondestructive testing: A succinct introduction. *Materials Evaluation*, **53** (4), pp. 461-462.

CHAPTER 2

Material characterization

2.1 INTRODUCTION

Material characterization is an important issue in many material production, processing and management applications. In production of materials that are composed of several constituents (mixtures), it is always important to keep the volume content of each constituent within a certain specification. Additionally, for materials in which a certain degree of chemical activity, such as curing, occurs (i.e. molecular linking of several substances), determination of the state of cure becomes an important issue as well. For example, when carbon loaded rubber is not properly cured, its mechanical properties are adversely influenced. Also, when concrete is not properly cured its compressive strength decreases. In composite materials lack of proper adhesive and binder cure can cause unforeseen internal defects, such as disbonds and delaminations while in use.

In the microwave and millimeter wave frequency regions it is possible to relate the electrical (and magnetic where applicable) properties of a material to those of its constituents and their volume contents. Furthermore, since curing process alters the properties of a material at its molecular level, it is possible to determine cure state in chemically processed materials as well. Also, by monitoring the electrical properties of the final product one is able to determine the presence of proper constituents and their volume contents, as well as monitoring the presence and level of any chemical activities.

Macroscopically speaking, the electrical properties of a material, which are of interest in this text, include its dielectric and conductivity characteristics, since the primary type of materials that are discussed here are insulators and conductors. However, the magnetic properties of materials are included in the discussions wherever necessary.

This chapter is prepared in such a way that those who are not familiar with the concept of dielectric properties could simply follow each section and obtain a clear understanding of the utilization of this parameter for material characterization. The goal of this chapter is not to provide the reader with a comprehensive and detailed understanding of the concept of dielectric materials (e.g. insulators), in particular as it relates to their understanding at an atomic level, and the procedure by which the dielectric constant is defined and derived. Thus, those interested in the subject of dielectrics at this level should consult the many references that are provided in this chapter, in particular (Balanis, 1989; Ramo, Whinnery and Van Duzer, 1994; von Hippel, 1954).

2.2 DIELECTRIC PROPERTIES

All matter is made of atoms in which several negatively charged electrons are held in various orbits around a positively charged nucleus. The electrons are held in these orbits as a result of the attractive forces in between the electrons and the nucleus (Wright and Skutt, 1965). Materials in which electrons in the most outer orbit, referred to as the valance band, are not tightly held are known as conductors. The electrons in the valance band of a conductor can freely transfer or jump from one atom to another. These are commonly referred to as free electrons. The number of these electrons is large in metals or conductors. In the absence of an impressed electric field, these electrons move in random directions and do not produce a net electrical current. In the presence of an applied static electric field these electrons freely move from one atom to the next, and solely contribute to the process of conduction in a conductor, and give rise to volume current density. For time varying fields the current density flows close to the surface of the conductor.

In dielectrics or insulators the electrons are tightly held in place by the forces between the electrons and the nucleus, and they are not free to move from one atom to the next (von Hippel, 1954). Therefore, they are known as bound charges. Consequently, there are no free charges in dielectrics that could contribute to the conduction process. When a dielectric is placed inside a static electric field the nucleus (positive charge) and the electrons (negative charges) of each of its atoms experience a force. This results in a relative change in the distance separating the positive and negative charges. Furthermore, all dipole moments align themselves with the electric field and produce a net dipole moment associated with the dielectric material, as shown in Fig. 2.1 (Balanis, 1989). Assuming a positive and a negative charge, q, associated with the electrons and the nucleus respectively, an electric dipole moment is produced and is given by (bold characters indicate vector quantities):

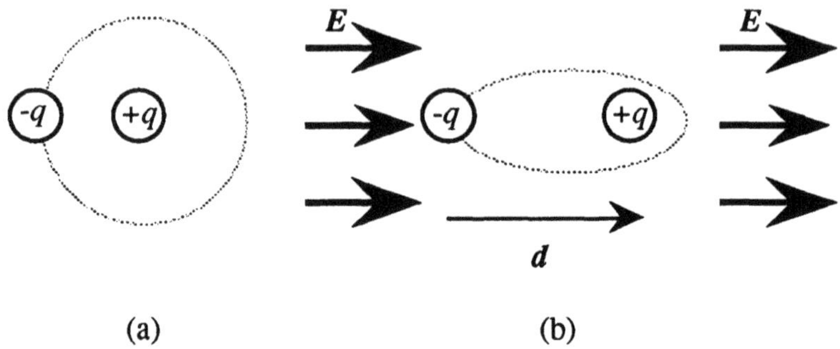

Fig. 2.1 Macroscopic representation of a dielectric atom, a) in the absence of a static electric field, b) in the presence of an impressed field.

$$\boldsymbol{p} = q\boldsymbol{d} \qquad (C \cdot m) \qquad (2.1)$$

The electric polarization vector associated with the dielectric is then the sum of all dipole moments on a per unit volume basis given by:

$$\boldsymbol{P} = \lim_{\Delta v \to 0} \left[\frac{1}{\Delta v} \sum_{i=1}^{n} \boldsymbol{p} \right] \qquad \left(C / m^2 \right) \qquad (2.2)$$

where n dipoles are present in a differential volume, Δv, and the unit C denotes Coulombs. Thus, the introduction of the polarization vector accounts for the generation of the electric dipole moments in the dielectric, and could be used to macroscopically describe the influence of an electric field on a bulk insulating material. In this way the generation and influences of each individual dipole moment are accounted for. This is accomplished by noting that the polarization vector is proportional to the impressed electric field by the electric susceptibility of the material. Finally, the electric flux density and the impressed electric field are related to each other by ε which is known as the dielectric constant (or for the static case the permittivity) of the dielectric material and is given by:

$$\varepsilon = \varepsilon_o (1 + \chi_e) \qquad (F / m) \qquad (2.3)$$

where ε_o is the permittivity of free-space (8.854×10^{12} F/m), χ_e is the electric susceptibility of the material and the unit F denotes Farads. Therefore, all that is needed to electrically understand the properties of the polarized material is its dielectric constant. This parameter macroscopically accounts for and expresses all of the changes at the atomic level as a result of polarization of the material by the impressed electric field.

The slight stretching of the atom as shown in Fig. 2.1(b) is similar to stretching a spring or lifting an object. There is potential energy associated with both of these events and such is true about the stretched atom. In effect, the dielectric composed of many such stretched atoms is said to have stored electrical energy. An example would be a capacitor employing a dielectric material and its capability to store electrical energy. There are three basic mechanism of polarization in dielectric materials (Balanis, 1989):

- **Electronic Polarization** - Most dielectrics exhibit this type of polarization in which, as explained above the centroids of the negative charges (electrons) and the nucleus (positive charge) in an atom experience a slight displacement in the presence of an electric field (Fig. 2.1).

- **Orientational (Dipole) Polarization** - Certain materials, known as polar materials, are composed of molecules that naturally posses a non-zero electric dipole moment, such being a water molecule. In the absence of an electric field, the dipole moments of these polar materials are randomly orientated resulting in no net dipole moment. However, in the presence of an electric field these dipole moments align and give rise to a net polarization vector.

- **Ionic (Molecular) Polarization** - Some materials consist of positive and negative ions. In the presence of an electric field a displacement in the relative location of these ions occurs which, similar to electronic polarization, gives rise to a net polarization vector.

Electrets are materials that posses a net polarization vector in the absence of an electric field. Ferroelectric materials are a group of dielectrics, that similar to ferromagnetic materials, exhibit hysteresis of polarization vector as a function of the impressed electric field (Rose, Shepard and Wulff, 1966).

For time varying electric fields, the polarization vector and consequently the dielectric and conductive (and magnetic) properties of a material are influenced by the alternating nature of the impressed field. These influences make for different dielectric properties as a function of frequency than those of the static case. The classical model of electronic polarization (Fig. 2.1) in which the stretching of the atoms produces a restoring force is similar to a classical mass-spring model (with friction) (Balanis, 1989; Ramo, Whinnery and Van Duzer, 1994). In this way the effect of frequency on the material is modeled by an equivalent spring coefficient, friction and mass (usually the electron mass in an atom shown in Fig. 2.1). In turn parameters such as damping and resonance may be easily incorporated into this model (also similar to an RLC circuit). For ionic polarization this resonance tends to occur at lower frequencies due to the heavier mass of the ions. Another way of envisioning the effect of frequency is the periodic changes of the polarity associated with the polarization vector. Since the nucleus of an atom is much heavier than the electrons orbiting it, the electron cloud primarily resonates back and forth around the nucleus. In the presence of friction in the system, and considering that numerous dipole moments are present, these alternating movements result in a certain amount of the impressed energy to be absorbed by the system. Therefore, for time varying cases the dielectric properties of a material must not only account for the way the material stores energy (such as that explained for the static case) but must also account for the way it absorbs energy. The absorbed energy is caused by the static conductivity of the material as well as the dielectric hysteresis effect. This effect is due to the alternating conductivity caused by the rotation of the polarization vector in an alternating electric field. Thus, the dielectric property of a material, under a time-varying condition, is given by (Kraus, 1992):

$$\varepsilon = \varepsilon' - j\varepsilon'' = \varepsilon' - j\left(\frac{\sigma_e}{\omega}\right) \tag{2.4}$$

where ε is the absolute complex dielectric property of the material, ε' is the absolute permittivity (related to displacement current) indicating the ability of the material to store energy, ε'' is the absolute loss factor (due to the static and the alternating conductivities) indicating the ability of the material to absorb energy, σ_e is the equivalent conductivity (accounting for both the static and the alternating conductivities) and ω is the radial frequency. For more detailed information regarding the effect of time varying electric fields on dielectric materials the reader is referred to the following references (Balanis, 1989; Ramo, Whinnery and Van Duzer, 1994; Kraus, 1992).

It is common to use the term 'dielectric constant' as opposed to the term 'dielectric property' as used here. The reason for using the latter term in this text is the fact that this property may not be constant as a function of frequency. Therefore, respecting this fact 'dielectric property' and 'dielectric properties' are used in this text. The relative to free-space complex dielectric property is then given by:

$$\varepsilon_r = \frac{\varepsilon}{\varepsilon_o} = \varepsilon'_r - j\varepsilon''_r \tag{2.5}$$

where ε'_r is referred to as the relative permittivity and ε''_r is referred to as the relative loss factor. The term loss tangent, ($\tan\delta$), refers to the ratio of the loss factor to permittivity. Materials with a loss tangent of zero are known to be lossless, while $\tan\delta << 1$ indicates a low-loss material and materials with $\tan\delta >> 1$ are considered to be high-loss materials.

The dielectric properties of a material made of several constituents (i.e. a dielectric mixture or a composite material) is dependent upon the dielectric properties of each of its constituents, their volume contents, their distribution in the mixture, the orientation of constituents with respect to the impressed electric field vector, any polymerization (i.e. molecular bonding and curing) that may have occurred during the production of the mixture and the operating frequency. It must be noted that it is also possible to use microwave energy as a heat source to promote and facilitate curing in various composite materials. Potentially, the following pieces of information may be obtained when evaluating the dielectric properties of a composite material or a mixture:

- dielectric properties of a constituent in the mixture,
- volume content of a constituent (i.e. porosity in polymers and ceramics),
- cure state,
- anisotropy associated with the mixture,
- physical and mechanical properties which may be dictated by the cure state of the material (i.e. compressive strength in concrete).

There are various theoretical and empirical dielectric mixing models that have also been developed in conjunction with this issue. Dielectric mixing models will be separately discussed in Section 2.7. There are also numerous methods for dielectric property measurement of materials. These techniques differ as a function of the following considerations:

- type of material to be evaluated (i.e. liquids, solids, gases, etc.),
- on-line or off-line requirements of the measurements,
- required measurement accuracy,
- type of measurement apparatus used such as open-ended transmission lines (e.g. waveguide, coax, open-resonators, etc.), cavity resonators, microstrip patches and filled (completely or partially) transmission lines,
- loss tangent of the material (i.e. low-loss vs. high-loss),
- destructive or non-destructive requirements of the measurement,
- in-contact or non-contact requirements of a measurement,
- geometry of the material under test (i.e. cylindrical fibers, sheets, etc.),

- particular piece of information that is sought after such as the volume content or the dielectric properties of the host (primary constituent of a mixture) or a given inclusion.

For a relatively complete and concise survey of microwave dielectric measurement techniques, the reader is referred to (Baker-Jarvis *et al.* 1995). In the next few sections, microwave dielectric evaluation of several mixtures, each having a different material characterization goal in mind, will be discussed. It must be noted that similar approaches may be undertaken for any other composite material of interest as well.

2.3 CARBON BLACK LOADED RUBBER[1]

Dielectric properties of rubber have been investigated during microwave curing process (Terselius, and Ranby, 1978; Schwarz *et al.*, 1973). The objectives of these studies have been to choose the optimum rubber composition as a function of microwave absorption (heating), and to show that microwave curing produces rubber compounds with similar properties as those produced by the standard curing process. Other investigators have reported tensor microwave dielectric property measurement of anisotropic rubber samples as it pertains to microwave shielding (Hashimoto and Shimizu, 1986; Sardos, Escarmant, and Saint-Christophe, 1990). Microwave techniques have also been used as diagnostic tools for the investigation of chemically reacting materials (i.e. epoxy) and studying polymerization process (Jow *et al.*, 1987; Martinelli, Rolla and Tombari, 1985). These studies show the potential of using microwaves for monitoring chemical reactions associated with the forming of cross-linked molecular networks in rubber compounds.

Very little is known about rubber dielectric properties as a function of microwave frequency. Also there is a lack of information about the behavior of rubber compound constituents (ethylene propylene diene rubbers or EPDM, oil, mineral filler, etc.) as a function of microwave frequency. Furthermore, there is very little to none reported on the influence of constituent volume fraction variation and chemical reactions among these constituents on the rubber dielectric property. These types of information may lead to important practical applications such as controlling rubber mixing process and introduction of new materials with predictable properties (i.e. microwave absorbers).

---

[1] Portions of this section are reprinted with permission from (Transactions on Microwave Theory and Techniques, MTT-42, no. 1, pp. 18-24, January 1994). © 1994 IEEE.

To study the dielectric properties of various rubber compounds as a function of microwave frequency one must first study the dielectric properties of rubber constituents; namely, EPDM, mineral filler, oil, zinc oxide and curatives. Subsequently, the influence of carbon black volume percentage on the dielectric properties of rubber must be investigated. The ability of microwaves to detect the presence of curatives in uncured rubber, and the role of frequency in this detection is presented here along with the impact of chemical reactions on the dielectric properties of rubber (Ganchev *et al.*, 1994).

2.3.1 Measurement procedure

There are many approaches for measuring the dielectric properties of materials at microwave frequencies. An extensive list of references as well as some useful practical comments are provided in the following references (Bussey, 1967; Baker-Jarvis *et al.* 1995). However, for studying the dielectric properties of rubber compounds in sheet form and their constituents the measurement apparatus must:

1. be able to measure low- and high-loss dielectric materials such as EPDM and carbon black loaded rubber with good accuracy,
2. be suitable for measuring powders (e.g. carbon black, zinc oxide), fluids (e.g. oils, curatives) and bulk materials (e.g. rubber sheets, EPDM),
3. be able to provide for measurement in a wide range of microwave frequencies,
4. be inexpensive and simple to use so that it could be reproduced easily, and be operated by those who are not trained in this area.

The well known completely-filled short-circuited waveguide technique (Roberts and von Hippel, 1946; Altschuler, 1963) was used which meets all of the above measurement requirements. This method is based on the measurement of the complex reflection coefficient of a load consisting of a dielectric sample completely occupying the cross-section of a rectangular waveguide while backed by a shorting plate. The magnitude and phase of the reflection coefficient are then used to calculate the dielectric properties of the sample filling the waveguide. Figure 2.2 shows the measurement apparatus that was used to measure the dielectric properties of rubber samples and their constituents. Five different waveguide set-ups operating in the frequency ranges of 3.95-5.85 GHz (G-band), 5.85-8.2 GHz (J-band), 8.2-12.4 GHz (X-band), 12.4-18 GHz (Ku-band), and 18-26.5 GHz (K-band) were used. The oscillator generates a microwave signal at the

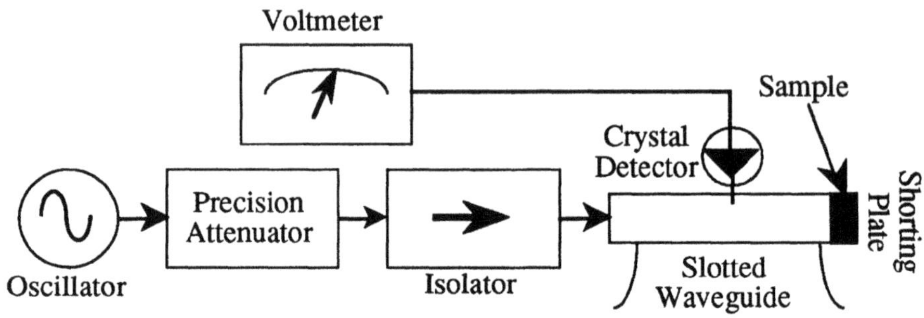

Fig. 2.2 Completely-filled short-circuited rectangular waveguide dielectric property measurement apparatus (Ganchev *et al.*, 1994).

desired frequency. This signal is then fed through a precision attenuator followed by an isolator into a slotted waveguide. The attenuator is used to eliminate the nonlinear characteristics of the detector, while the isolator protects the oscillator from strong reflections which may enter it. The standing wave characteristics inside the waveguide are measured via a detector and a sensitive voltmeter. The measurements were performed using precisely machined sample holders to accommodate samples of different thicknesses. For liquid and powder constituents the sample holder was filled completely and a piece of clear tape was used to hold them in place. It must be noted that a state-of-the-art microwave vector network analyzer, such as HP8510 or HP8720 series, is capable of accommodating these measurement. However, the cost associated with such a test equipment may be prohibitive.

The relationship between the measured complex reflection coefficient, Γ, of the sample backed by a shorting plate in the waveguide and its complex relative dielectric property, ε_r, is given by:

$$\varepsilon_r = \frac{\left(\frac{x}{kL}\right)^2 + \left(\frac{\lambda_g}{2a}\right)^2}{1+\left(\frac{\lambda_g}{2a}\right)^2} \qquad (2.6)$$

where x is a solution of the following transcendental equation:

$$\frac{\tan x}{x} = \frac{1-\Gamma}{jkL(1+\Gamma)} \qquad (2.7)$$

and k is the wavenumber inside the waveguide given by $2\pi/\lambda_g$. λ_g is the waveguide wavelength, L is the sample thickness and a is the broad dimension of the waveguide. It is known that equation (2.6) has an infinite set of complex roots (Altschuler, 1963). Thus, to find the correct root one must either have some idea of the value of ε_r, or make two or more measurements for different sample thicknesses. Hence, not only for obtaining the correct root, but also for increased measurement accuracy, Γ may be measured for several samples with different thicknesses. Γ as a function of ε_r and L can be expressed as:

$$\Gamma(\varepsilon_r, L) = \frac{\sqrt{\zeta} - j\tanh\left(kL\sqrt{\zeta}\right)}{\sqrt{\zeta} + j\tanh\left(kL\sqrt{\zeta}\right)} \tag{2.8}$$

where ζ is given by:

$$\zeta = \left[1 + \left(\frac{\lambda_g}{2a}\right)^2\right]\varepsilon_r - \left(\frac{\lambda_g}{2a}\right)^2 \tag{2.9}$$

From the measurements an array of values is obtained for $\Gamma(\varepsilon_r, L_n)$ where n is the number of samples with different thicknesses. The unknown dielectric property, ε_r, is then determined by finding the best fit to equation (2.8). The two measured parameters are:

$$SWR = 20\log(VSWR) = 20\log\left(\frac{1+|\Gamma|}{1-|\Gamma|}\right) \tag{2.10}$$

and the position of the standing wave null (which is related to the phase of Γ) in the slotted waveguide. The discrete points shown in Figs. 2.3 and 2.4 represent typical results for two rubber samples (the exact materials used to obtain these figures will be discussed later) of various sample thicknesses at 5 GHz. These values are then used to calculate the dielectric property of the rubber samples using the best fit to equation (2.8). The lines represent the results obtained from equation (2.8) using the calculated ε_r. It is important to note that more sample thicknesses render more accurate results. Additional improvements in the measurement accuracy can be obtained by using samples whose thicknesses are around the phase transition (minimum SWR) as shown in Figs. 2.3 and 2.4. The position of the phase transition may be calculated either from prior approximate knowledge of the dielectric constant or a preliminary measurement with a random sample thickness.

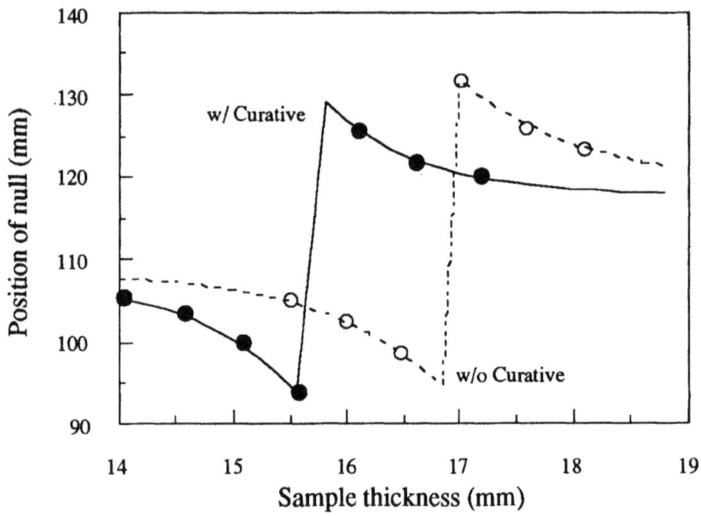

Fig. 2.3 Position of null vs. sample thickness at 5 GHz for two uncured rubber compounds with and without curatives (Ganchev *et al.*, 1994).

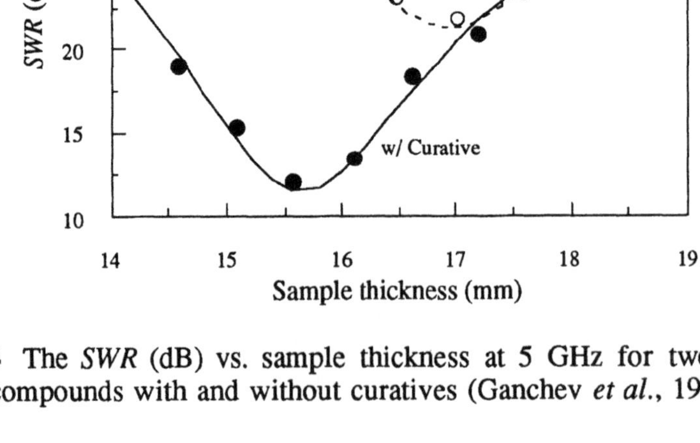

Fig. 2.4 The *SWR* (dB) vs. sample thickness at 5 GHz for two uncured rubber compounds with and without curatives (Ganchev *et al.*, 1994).

2.3.2 Dielectric properties of rubber compound constituents

To fully understand the dielectric properties of various rubber compounds (or any mixture), one must first have a complete knowledge of the dielectric properties of each of their individual constituents which in the case of rubber include EPDM, mineral filler, zinc oxide, oil and curatives. Table 2.1 lists the dielectric properties of the primary constituents of a rubber compound. The respective volume percentages shown are used for a rubber sample that is considered the basic sample throughout this study. Due to the very high conductivity of carbon black, its dielectric properties are not measured. This table shows the mean value of relative permittivity and loss tangent of the constituents measured at several frequencies in the X-band frequency region (8.2-12.4 GHz). Measurements in other frequency bands resulted in similar values of dielectric properties for these constituents. This is primarily due to the fact that all of these constituents are shown to be low-loss materials. The uncertainty of these measurements is relatively higher than (about 5% for ε'_r and 10% for $\tan\delta$) all other measurements reported later in this section, since only two random thicknesses of each constituent were used in these measurements. The measurement accuracy associated with the measurement approach used here is discussed in Section 2.3.5. The results indicate that the dielectric properties of EPDM, oil and mineral filler are for all practical purposes equal. This is true in light of the fact that rubber and carbon black dielectric properties are much higher than these values. Zinc oxide and curatives have slightly higher dielectric properties but they have small volume percentages in the total composition of a typical rubber compound. Through only physical mixing (not considering any chemical reaction), the contribution of curatives to the rubber compound dielectric property should be negligible. However, as will be seen later the chemical reaction triggered by the curatives in uncured rubber samples causes a detectable change in the dielectric properties of rubber at ambient temperature.

Table 2.1 Basic rubber sample primary constituents volume content (%) and their X-band dielectric properties (Ganchev *et al.*, 1994).

Constituent	*Volume %*	ε'_r	$\tan\delta$
EPDM	45.50	2.1	0.008
Oil	25.85	2.2	0.014
Carbon black	17.00		
Mineral filler	9.30	2.3	0.017
Curatives	1.80	3.1	0.032
Zinc oxide	0.36	3.9	0.280

2.3.3 Cured rubber dielectric property dependence on carbon black

In the production of rubber one of the most important issues is the influence of carbon black volume content (%) on the dielectric properties of rubber compounds. Consequently, measurements were performed on fourteen rubber samples with varying degrees of carbon black concentration. The rubber compounds were prepared by carefully measuring the ingredients in a precision laboratory scale. They were loaded in a laboratory Banbury mixer by upside down method and mixed for five minutes at maximum temperature of 130^{oC}. Then, the compounds were mixed for another five minutes at ambient temperature in an open two-roll mill. The rubber sheets were vulcanized in molds at 150^{oC} for a predetermined time. The carbon black used for this study was aggregate of particles ranging in diameter from 5 to 125 nanometers. During the mixing process the aggregates are usually broken in smaller sizes and distributed uniformly throughout the mass. The rubber samples were prepared starting from the basic formulation given in Table 2.1. Then, the volume percentage of carbon black was varied between 8.40 to 35.61. This was done in such a way that any change in the amount of carbon black percentage is at the expense of EPDM, while the percentages of all other constituents are kept constant (same percentages as shown in Table 2.1). The goal is to isolate the influence of carbon black on rubber compound dielectric properties. Table 2.2 lists the carbon black volume content for the measured rubber samples (sample 5 is the basic sample).

Table 2.2 Carbon black volume content (%) for measured samples (Ganchev *et al.*, 1994).

Sample	*Volume content (%)*
1	8.40
2	10.13
3	12.13
4	14.41
5 (basic)	17.00
6	18.02
7	19.07
8	20.14
9	21.23
10	22.34
11	25.29
12	28.58
13	32.05
14	35.61

These samples were prepared in sheet forms with thicknesses of 0.5, 1, 2, 3 and 5 mm to accommodate the completely-filled short-circuited waveguide technique, and their dielectric properties were measured at 5, 7, 10, 16 and 24 GHz. Figures 2.5 and 2.6 show the measured relative permittivity and loss tangent at 5 and 24 GHz, respectively (results for other frequencies fall in between these two extremes). The results show that the dielectric properties of rubber increase as the carbon black volume percentage increases. Clearly, dielectric property variations are more pronounced at 5 GHz which render this frequency (or lower) useful for dielectric constant variation measurement of these rubber compounds.

It has been reported that the rolling effect associated with the production of rubber sheets causes anisotropy (Hashimoto and Shimizu, 1986). To alleviate this problem, these samples were prepared in such a way that rolling effect was minimized by performing the rolling process in many different directions. To check the success of this manipulation, the dielectric properties of several samples for several arbitrarily oriented (with respect to the electric field vector in the waveguide) pieces were measured. The results indicated no systematic deviations in dielectric property values (i.e. within the measurement errors) which shows the absence of anisotropy in these samples.

2.3.4 Detection of curatives in uncured rubber

The presence of curatives was checked for two types of rubber; namely, carbon black loaded and mineral loaded rubber. The goal of these experiments was to investigate the possibility of using microwaves to detect the presence (or absence) of curatives prior to product making step and curing. All of the following measurements were performed at ambient temperature.

(a) Curatives in carbon black loaded rubber

Uncured version of samples 4, 5, 7, 9, 11 and 14 (referring to Table 2.2) were prepared with and without curatives. These samples were chosen to represent the entire range of carbon black percentages used in most rubber products. It was soon determined that the dielectric property variations for these samples were most pronounced at 5 GHz. It was also learned that the dielectric property difference between these two types of rubber samples is small. Therefore, as already discussed, to distinguish between these two types of samples, the sample thicknesses used for dielectric property measurement are best to be around the phase transition regions. Figures 2.3 and 2.4 show typical results of such measurements (sample 5 at 5 GHz).

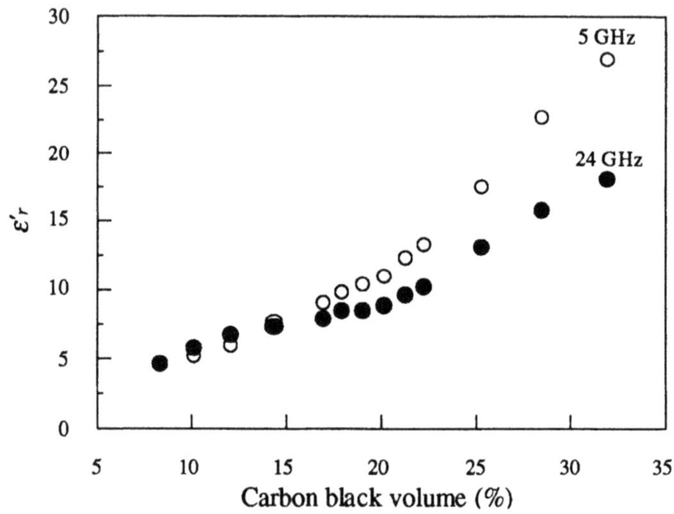

Fig. 2.5 ε'_r for all cured rubber compounds at 5 and 24 GHz (Ganchev *et al.*, 1994).

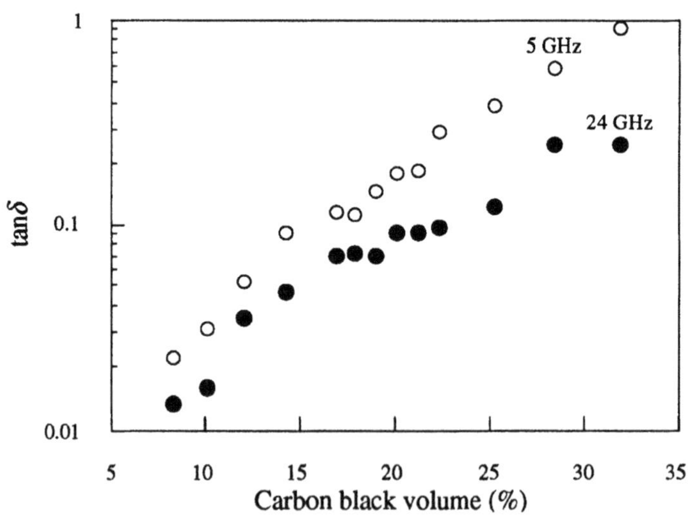

Fig. 2.6 tanδ for all cured rubber compounds at 5 and 24 GHz (Ganchev *et al.*, 1994).

It is obvious that for sample thicknesses between 15 mm and 16 mm a *SWR* difference of more than 10 dB is measured. The opposite is true for measurements between 17 mm and 18 mm. These examples illustrate the importance of choosing optimum sample thicknesses to detect minute dielectric property differences for a given frequency.

Figures 2.7 and 2.8 show the relative permittivity and the loss tangent for uncured samples with and without curatives at 5 GHz as a function of carbon black volume percentage. From Fig. 2.7 it is apparent that the addition of curatives does not affect the relative permittivity of the rubber compound in a uniform fashion for all carbon black volume percentages. However, in spite of the low losses exhibited by curatives themselves (Table 2.1), the loss factor of these rubber samples tends to increase with the addition of curatives for all carbon black volume percentages. Furthermore, in the presence and absence of curatives, the difference in $\tan\delta$ is consistently detected as shown in Fig. 2.8. The detected difference is attributed to a chemical reaction triggered by the curatives at room temperature. It is well known, in the rubber industry, that a rubber compound containing curatives will cross link (cure) in due time even at room temperature. The rate of curing will depend upon the type and the amount of curatives, the polymer and the carbon black. It is apparent from these measurements that the initial stage of curing can be detected. As it was discussed earlier, the difference in the dielectric properties of a rubber compound in the presence of curatives can not be described by a simple physical mixing effect (curatives occupy only 1.8% of the total rubber volume with dielectric properties close to those of EPDM). Another observation is that after the carbon black percentage increases beyond 25%, the difference between the samples with and without curatives becomes smaller. It may be hypothesized that at these carbon black percentages the effect of the chemical reaction caused by the presence of the curatives is masked by the overwhelming presence of unlinked carbon black.

(b) Relationship among EPDM, carbon black and curatives

The measurements thus far involved a constant percentage of curatives while the ratio of carbon black to EPDM was changed. This may not reflect the composition of a real rubber product. Subsequently, having noticed the effect of curatives on dielectric properties, a new sample (named sample 15) was prepared. The dielectric properties of this sample when cured and not cured (with and without curatives) were then measured. Based on the cured results it was predicted that the carbon black content is close to that of sample 11 (25%). The measured dielectric properties of both samples are listed in Table 2.3. However, a glance at the uncured compound compositions, listed in Table 2.4,

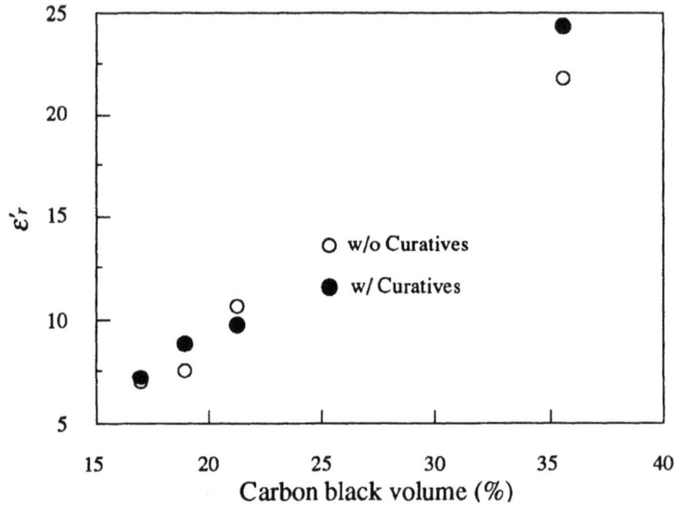

Fig. 2.7 ε'_r for uncured rubber compounds with and without curatives at 5 GHz (Ganchev *et al.*, 1994).

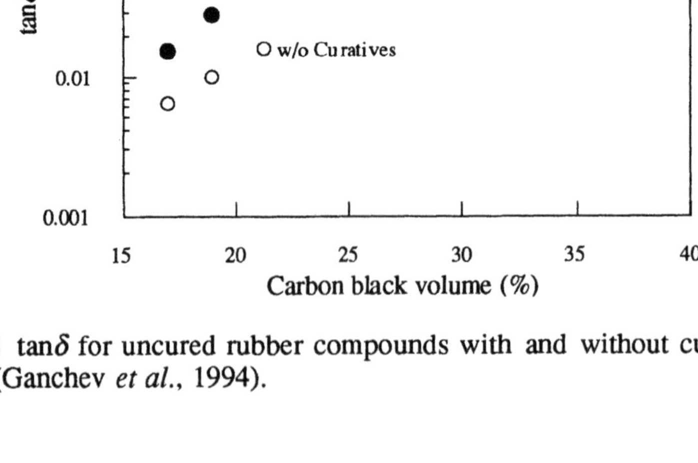

Fig. 2.8 tanδ for uncured rubber compounds with and without curatives at 5 GHz (Ganchev *et al.*, 1994).

Table 2.3 Measured dielectric properties of rubber samples 15 and 11 for cured and uncured with and without curatives (Ganchev *et al.*, 1994).

	Sample number 15		*Sample number 11*	
	ε'_r	$\tan\delta$	ε'_r	$\tan\delta$
Cured	18.8	0.400	17.4	0.385
Uncured w/o	17.6	0.099	13.6	0.046
Uncured w/	16.0	0.056	11.6	0.062

Table 2.4 Primary constituent formulation of samples 15 and 11 (Ganchev *et al.*, 1994).

Constituent	*Sample 15 volume (%)*	*Sample 11 volume (%)*
EPDM	22.0	34.84
Carbon black	33.8	25.30
Oil	32.4	27.50
Mineral Filler	7.53	9.85
Curatives	0.85	1.91
Zinc oxide	0.29	0.28

indicates that samples 15 and 11 are not similar which may suggest the following. It is apparent that the volume percentage of carbon black in sample 15 is higher than that in sample 11. This is confirmed only if we focus on the dielectric properties of the uncured samples. Then, if the uncured dielectric property values for sample 15 and Figs. 2.7 and 2.8 are used, we may closely predict the carbon black percentage. Thus, for the samples without curatives only a physical mixing process may be considered. However, once cured, the chemical reaction will have changed the properties of the rubber compound, and hence its dielectric properties.

The volume percentages of EPDM and curatives are considerably higher in sample 11 than in sample 15. Therefore, it may be hypothesized that similarity between the dielectric properties of the cured samples is due to the fact that the lack of carbon black in sample 11 is compensated by the increased reaction involving EPDM. What is not expected is the importance of this reaction (cross-linking) which in this case compensates for the influence of carbon black on the dielectric properties. There still remains a question regarding the reduction in loss tangent, for uncured with and without curatives, in the opposite directions between sample 15 and sample 11. Obviously, here some phenomenon which decreases the losses, and may be due to formation of cross-linked molecular network in such a way that dipoles rotate less freely (Jow *et al.*, 1987) is encountered. Most likely there are two opposite mechanisms influencing the dielectric properties of uncured rubber when curatives are added depending on the carbon black, EPDM and curatives volume percentages. Clearly, further studies are needed to fully understand the impact of these mechanisms on the dielectric properties.

(c) Effect of curatives in mineral loaded rubber

Similar measurements were performed on uncured samples with and without curatives in which carbon black was replaced by a mineral filler. As expected, this sample exhibits different dielectric properties than its carbon black counterpart. The results for this measurement are presented in Table 2.5. From this table one may conclude that for such a low-loss rubber, the presence of curatives is difficult to be detected. The difference between the two samples is comparable with the measurement accuracy reported in the next section. However, it is believed that this difference may still be detected if samples with optimum thicknesses are used. For thicknesses just after the second phase transition (between 29 mm and 32 mm) at 5 GHz a repeatable difference of about 3 dB in the value of *SWR* was detected. This difference will be more pronounced for thicknesses after the third phase transition. This part of the study confirms that the effect of curatives on carbon black loaded rubber dielectric properties may not be limited to a chemical interaction involving EPDM (cross-linking), but carbon also plays a significant role in this chemical process.

Table 2.5 Dielectric properties of uncured mineral loaded rubber with and without curatives (Ganchev *et al.*, 1994).

Uncured sample	ε'_r	$\tan\delta$
Without curatives	3.00	0.0090
With curatives	2.91	0.0107

2.3.5 Measurement accuracy

The measurement uncertainty associated with the relative permittivity measurements using transmission line method is reported to be about 5% and 10% for loss tangent (Baker-Jarvis *et al.* 1995; Chao, 1986). The accuracy of the measurement technique used in this study is well established (Tsankov, 1981; Chao, 1986). However, the measurement accuracy has been improved by measuring multiple sample thicknesses around the phase transition region and direct fitting of the measurement results to obtain the dielectric properties. This is the reason for the consistent ability to detect curatives in uncured rubber samples particularly for low-loss materials.

The measurement technique reported here consisted of determining two parameters (*SWR* and relative null position) for various sample thicknesses. The accuracy of these measurements are increased if a rotary-vane attenuator and a precision slotted line is used (as was done here). For high *SWR* cases the following correction factor was used:

$$SWR_c = -20\log\left(10^{\frac{-SWR}{20}} - 10^{\frac{-SWR_o}{20}}\right) \qquad (dB) \qquad (2.11)$$

where SWR_c is the corrected value of the parameter shown in equation (2.11). *SWR* in (dB), is the measured value, and SWR_o in (dB), is the value for the short-circuited waveguide. The use of this correction factor is justified because a lossless waveguide and a perfect short circuit termination were assumed. Equation (2.11) shows that there is a limitation in the measurement of low-loss materials (*SWR* approximately equal to SWR_o). However, the use of thicker samples of low-loss materials can reduce this problem.

Specific precautions must be considered for low- and high-loss materials. For low-loss materials the relative permittivity may be precisely obtained from the multiple thickness null location information, but the relative loss factor may not be deduced with reasonable accuracy. Thus, for this case the *SWR* measurement is very important as this is the sole indication of losses. For high-loss materials a different problem exists. The null location curve quickly degenerates to a sloped line as a function of sample thickness. This is why the precise value of dielectric properties can not be extracted from the null location measurements alone. In this case the *SWR* measurement plays a crucial role as well. Thus, for high-loss materials, measurements should be performed close to the first *SWR* minimum, because beyond that the *SWR* oscillates negligibly as a function of sample thickness.

It is difficult to estimate the measurement accuracy for this technique directly, since a straightforward approach in determining the dielectric properties of each sample (see equation (2.8)) was not used. However, from the repeatability of the results, especially when detecting small changes in losses due to the presence of curatives, it is clear that the accuracy is better than the usually accepted for a waveguide method (Tsankov, 1981; Chao, 1986). Consequently, a numerical approach may be used to estimate the apparatus measurement error. The following sources of error were considered: uncertainties due to the resolution error of the rotary-vane attenuator (±0.25 dB), the position of the standing wave minimum (±0.05 mm), and the thickness of the sample (±0.1 mm). The error due to the frequency instability associated with the oscillator may be neglected if a stable oscillator is used. The analysis consisted of the following steps. The uncertainty associated with the measured quantities for all sample thickness was added (or subtracted). Then dielectric property, and subsequently the percentage difference for the relative permittivity and the loss tangent were calculated. Finally, the maximum error (worst case) for a given measurement was chosen. This procedure was performed for all the measurements reported in previous section. This resulted in a measurement uncertainty of about 1% for the relative permittivity and about 3.5% for the loss tangent. The uncertainty is worse if measurement points are not in the vicinity of the phase transition.

2.4 RESIN BINDER[2]

Low density fiberglass composites are used in many environments for insulation purposes. There are several important issues associated with the production of fiberglass products; namely:

- the uniformity by which the resin binder is applied,
- resin binder cure state,
- variations in the glass fiber density.

In this section, the results of dielectric property evaluation of fresh and 12-day old resin binder (indicating small amount of curing), resin binder loaded but uncured fiberglass, fiberglass with no resin binder and cured fiberglass with three different resin binder levels in the frequency range of 4-18 GHz, using the same completely-filled short-circuited waveguide approach, are reported (Qaddoumi, Ganchev and Zoughi, 1996; Roberts and von Hippel, 1946; Ganchev *et al.*, 1994). To measure the dielectric properties of the liquid binder an upright short-circuited waveguide was used, as shown in Fig. 2.9. In this way, measurements using many sample thicknesses (in this case lengths) of the liquid resin binder could be made.

2.4.1 Dielectric properties of fresh and 12-day-old resin binder

The dielectric properties of fresh liquid resin binder and resin binder left at room temperature for 12 days were measured in a wide frequency range of 4-18 GHz. The resin binder is usually stored in a freezer since if left at room temperature, for even a short time, its properties will sufficiently change, and it will not be used in fiberglass production. This change is attributed to the occurrence of a certain level of curing at room temperature. Consequently, having been left at room temperature for 12 days, it is expected that the resin binder has been somewhat cured.

[2] Portions of this section are reproduced with permission from (Research in Nondestructive Evaluation, Microwave Diagnosis of Low Density Glass Fibers with Resin Binder, Qaddoumi, Ganchev and Zoughi, vol. 8, no. 3, pp. 177-88) © 1996 Springer-Verlag.

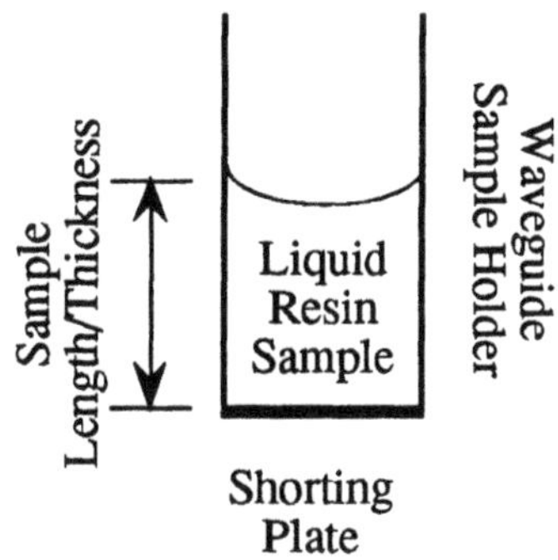

Fig. 2.9 Side-view of an upright completely filled short-circuited waveguide with a dielectric fluid (Qaddoumi, Ganchev and Zoughi, 1996).

Therefore, any measured difference in the dielectric properties of fresh and 12-day-old resin binder is expected to be due to this curing. Figures 2.10 and 2.11 show the measured and calculated results of the position of null and *SWR* at 6 GHz as a function of the sample thicknesses for both resin binders. The dielectric properties are once again found by the best fit method from the measured data using equations 2.8 and 2.9.

The apparent differences between the measured and the calculated results, shown in these figures, are due to the following:

- the resin binder (mostly water) is a high-permittivity and high-loss dielectric material presenting some difficulties when measuring its dielectric properties (refer to the discussion in section 2.3.5),
- the resin binder is a liquid and when filling the sample holder a meniscus is produced as depicted in Fig. 2.9. This results in a certain error associated with measuring the sample length which in this case is calculated from the resin binder volume. In the measurements this effect was accounted for by using an empirically established effective sample length when calculating the dielectric properties.

The goal of these measurements is not to obtain the precise value of the relative permittivity and loss factor of these samples, but rather to look for any differences between these values for the fresh and the 12-day-old resin binders. Figures 2.10 and 2.11 demonstrate that at some sample thicknesses, there is a substantial difference between the measured parameters (e.g. 4 mm in Fig. 2.10 and 2.5 mm in Fig. 2.11). Conversely, at other sample thicknesses (e.g. 1.5-2.5 mm in Fig. 2.10 and 3.5-3.7 mm in Fig. 2.11) the measured differences are negligible. This also demonstrates the effectiveness of multiple sample thicknesses measurement as opposed to arbitrarily choosing a sample thickness.

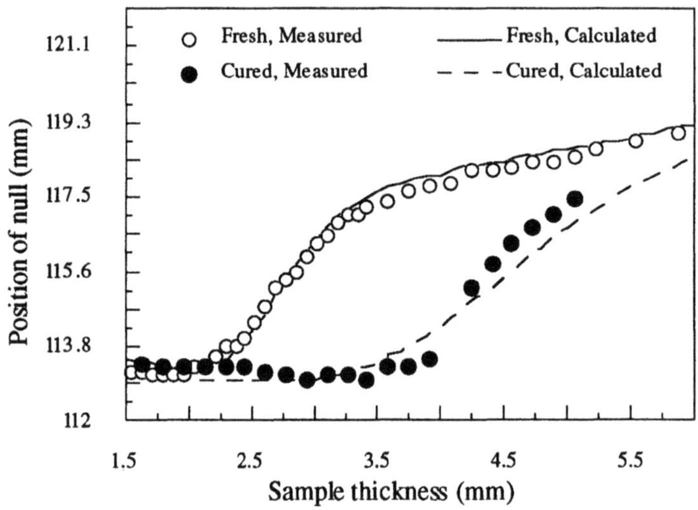

Fig. 2.10 Position of null vs. sample thickness at 6 GHz for fresh and 12-day-old resin binder (Qaddoumi, Ganchev and Zoughi, 1996).

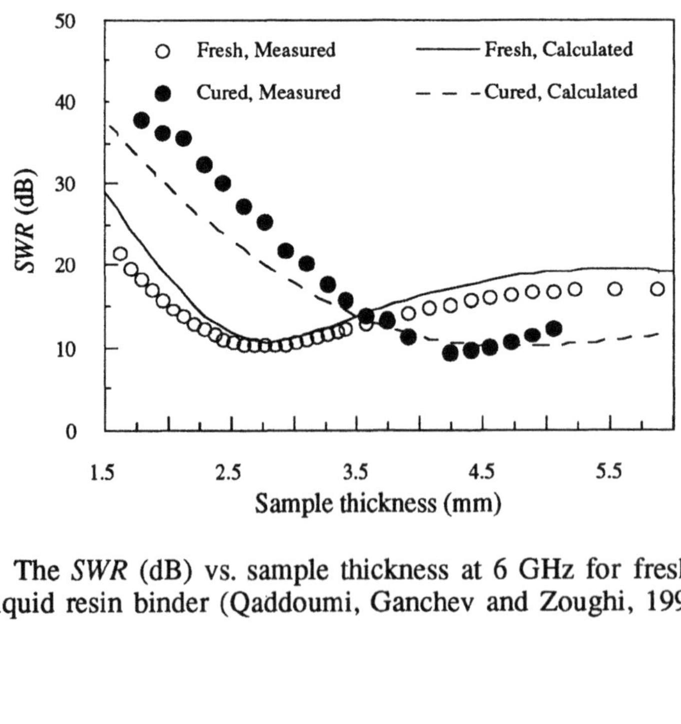

Fig. 2.11 The *SWR* (dB) vs. sample thickness at 6 GHz for fresh and 12-day-old liquid resin binder (Qaddoumi, Ganchev and Zoughi, 1996).

Additionally, these results illustrate that in practice the standing wave properties alone may be used to distinguish between these two resin binders. This is an attractive practical feature since standing wave measurements are usually conducted with relatively simple and inexpensive microwave measurement systems. Using the measurement results shown in Figs. 2.10 and 2.11, the dielectric properties of these two resin binders are calculated using equations 2.8 and 2.9. Table 2.6 shows the results of these measurements as a function of frequency. The results indicate that the difference between the relative permittivity of these resin binders is frequency dependent. Additionally, at most frequencies this difference is considerable. The percentage difference between the relative permittivities of these two samples (with respect to the fresh resin binder relative permittivity) is also listed in Table 2.6. Thus, this information alone may be used to determine the presence and initiation of the curing process. Loss tangent results are close to each other, and may not be suitable for distinction between these two resin binder samples. The measurement results listed in Table 2.6 clearly indicate that for a relatively slight amount of curing occurred at room temperature, the dielectric properties of the resin binder (particularly its relative permittivity) is sufficiently and substantially influenced. Consequently, one would expect a larger change in the dielectric properties of the resin binder when a significant amount of curing has taken place. Hence, these preliminary measurements demonstrate the potential capabilities of microwave dielectric property measurements for cure state monitoring in these materials.

Table 2.6 ε'_r and $\tan\delta$ for fresh and 12-day-old resin binder samples vs. frequency (Qaddoumi, Ganchev and Zoughi, 1996).

Frequency (GHz)	*Fresh binder* ε'_r	*Fresh binder* $\tan\delta$	*12-day-old binder* ε'_r	*12-day-old binder* $\tan\delta$	$\Delta\lvert\varepsilon'_r\rvert\%$
4	21.0	0.79	22.5	0.79	7.14
4.5	18.8	0.76	20.5	0.74	9.04
5	17.2	0.83	19.4	0.84	12.79
5.5	16.0	0.81	18.7	0.88	16.88
6	15.1	0.80	16.4	0.69	8.61
8.5	12.3	0.87	11.8	0.83	4.07
10	10.3	0.99	9.8	0.95	4.85
12	8.6	1.18	7.2	1.27	16.28

2.4.2 Resin Loaded Fiberglass

The dielectric properties of fiberglass samples with resin binder levels (by weight) of 0%, 9.4%, 13.8% and 18.6% along with a sample with resin binder but uncured were measured at 10 GHz as well. Figures 2.12 and 2.13 show the measured and calculated results of the position of null and *SWR* as a function of sample thickness for the sample with uncured resin binder at 10 GHz. The agreement between the measured and the calculated results is better than those for the liquid resin binder, but still this agreement is not as good as those shown in Figs. 2.3 and 2.4. This is due to the fact that the fiberglass material was compressed in the waveguide sample holder. There is a limited control over how well one is able to push out the air in between the fibers and how much expansion after compression occurs which result in slight error when determining the sample length.

Table 2.7 shows the measured dielectric properties for all fiberglass samples at 10 GHz. For these samples, the relative permittivities have similar values, but loss tangent values are different. However, it may still be argued that it is difficult to use dielectric property information to distinguish among these samples. Therefore, to distinguish among these fiberglass samples a different approach must be used. Later, in Chapter 5 which deals with microwave near-field imaging, it will be shown that an open-ended rectangular waveguide probe (an optimized measurement) is easily capable of detecting these resin binder level variations (Qaddoumi, Ganchev and Zoughi, 1996).

Table 2.7 ε'_r and $\tan\delta$ for five different fiberglass samples at 10 GHz (Qaddoumi, Ganchev and Zoughi, 1996).

Fiberglass	ε'_r	$\tan\delta \times 10^3$
No resin binder	1.80	7.1
w/ uncured resin binder	1.70	9.2
w/ 9.4% resin binder	1.55	6.5
w/ 13.8% resin binder	1.50	11.2
w/ 18.4% resin binder	1.60	13.8

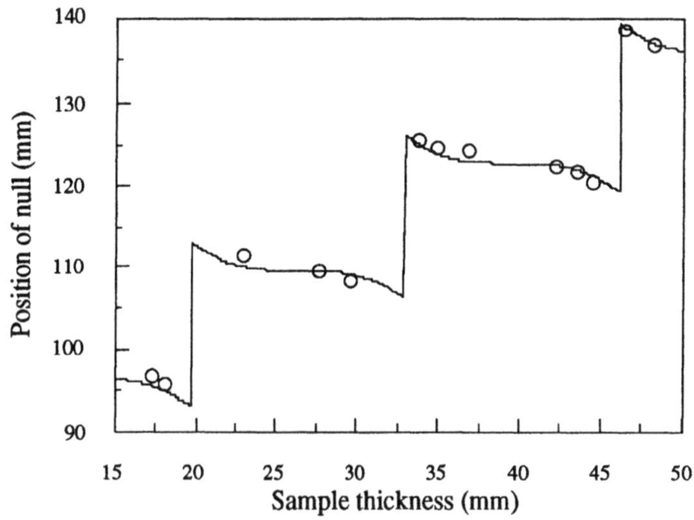

Fig. 2.12 Position of null vs. sample thickness at 10 GHz for uncured fiberglass with resin binder (Qaddoumi, Ganchev and Zoughi, 1996).

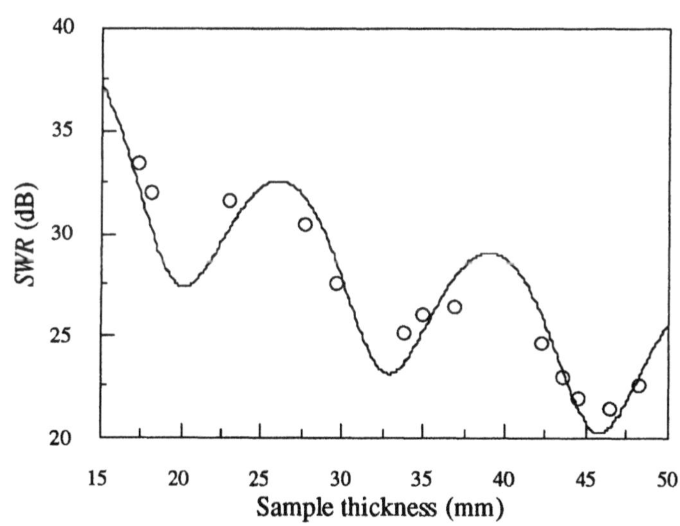

Fig. 2.13 The *SWR* (dB) vs. sample thickness at 10 GHz for uncured fiberglass with resin binder (Qaddoumi, Ganchev and Zoughi, 1996).

2.5 POROSITY ESTIMATION IN POLYMER COMPOSITES[3]

Determining porosity level (or the amount of air content) in cured polymers, ceramics and composite materials is an important practical issue. In cured polymers, the presence of porosity causes lowered mechanical performance due to stress concentrations. Localized porosity can be particularly damaging to the joint strength of adhesively bonded components. In ceramics, the relative density is an important processing parameter, and again the ceramic is extremely sensitive to stress concentration (lowered density). If not fully densified, a ceramic is weak and has low stiffness. In composites, the porosity can be within the matrix material which will affect the performance in a similar fashion to those in bulk materials. However, porosity often concentrates at specific locations in composite materials (either between plies or at the fiber/matrix interface), and can dramatically lower flexural and shear performance. Increases in porosity during operation (material under loading) may precede macroscopic damage and possibly indicate the presence of delamination. Hence, a technique capable of detecting and accurately determining porosity level in materials is desirable.

Several techniques are available for determining porosity in polymer matrix composites. Of these, ultrasonics techniques are the most frequently used. They are based on the measured amplitude attenuation of an ultrasonic longitudinal wave pulse. Attenuation can either be measured through the material, as in the through-transmission technique, or as reflected off the back surface of the material as in the pulse-echo technique. Ultrasonic techniques commonly require the use of a coupling agent (such as water or oil) between the material and the ultrasonic transducer. This is a disadvantage when inspecting polymeric composites since moisture uptake degrades their mechanical performance (Daniel and Komsky, 1992; Jones and Berger, 1989; Rubin and Jerina, 1993; Thompson *et al.*, 1991). Radiography techniques are also used to image composite porosity. The absorption and scattering of X-rays by different materials allows the detection of porosity. The technique described by Lipowitz *et al.*, 1990, offers very accurate determination of void size down into nanometer scale but the image is 2-D. Only through thickness, porosity can be measured using this technique, and small scale porosity detection may not be possible when blocked by larger objects. These limitations may be overcome through the use of computed tomography X-ray imaging, but at a relatively great expense. Using this technique 3-D images are created with good contrast resolution (Armistead and Yancey,

---

[3] Portions of this section are reprinted with permission from *Materials Evaluation*, vol. 53, no. 3, pp. 404-08, Copyright 1995 © The American Society for Nondestructive Testing, Inc.

1989). Additionally, real-time imaging is not possible with this technique. Video processing of the X-ray image is available but system costs limit the use of this techniques (Jones and Berger, 1989; Lipowitz *et al.*, 1990; Armistead and Yancey, 1989; Smith, Hua and Earl, 1994). Finally, neutron radiography has also been used to image composite porosity as well as moisture uptake. Unlike X-rays, attenuation of thermally activated neutrons is not linearly dependent on the target materials atomic weight. Neutrons are preferentially attenuated by certain lighter elements such as Hydrogen and Boron. Hence this technique is particularly useful for water detection. Equipment size and weight limit this technique to testing centers capable of acquiring such equipment (Jones and Berger, 1989; Smith, Hua and Earl, 1994).

Microwave techniques possess certain advantages over the other approaches mentioned. In particular, microwave techniques are well suited for inspecting polymers and ceramics. This is partially due to the fact that microwaves easily penetrate inside dielectric material and the reflected or transmitted signal can be related to the dielectric properties of the material, and subsequently to its physical and mechanical characteristics.

In this section the potential of microwave dielectric property characterization for porosity determination in low-permittivity and low-loss composites samples is demonstrated. This information can then be used for optimization of several established non-destructive microwave techniques to determine porosity levels in composites (Bakhtiari, Ganchev and Zoughi, 1993 ; Bakhtiari *et al.*, 1994).

2.5.1 Sample preparation

Samples of polymer microballoon-filled epoxy resin with 0%, 48.9%, 58.7% and 68.5% air volume fractions were carefully prepared (Gray *et al.*, 1995). The uniformly distributed air-filled microballoon sizes ranged, in diameter, between 15 to 200 micrometers with a bulk density of 0.009 gm/cm^3. These microballoons created the desired porosity levels in these samples. The samples were shaped to fit inside short-circuited rectangular waveguides (see section 2.3.1) and the measurements were conducted at frequencies of 8.2, 10, 12, 14, 16 and 18 GHz. Long polymer air-filled microballoon samples were made in such a way that their cross-sections were rectangular and tightly fit the waveguide sample holders. The thicknesses of these samples were then cut down as the experiments proceeded to provide multiple sample thicknesses required in these measurements. The goal of these measurements is to study the sensitivity of microwaves to small changes in porosity level (as it relates to changes in dielectric properties). This study was conducted so that based on its results we may extrapolate the potential utility of microwave techniques in more complicated composites such as polymer matrix composites.

Figures 2.14 and 2.15 show the variation of measured null position and *SWR* as a function of sample thickness at 10 GHz (for the sample with 0% porosity), respectively. The measured values (circles) are once again used to calculate the dielectric properties of the sample using equations 2.8 and 2.9, as done in the previous cases. The lines represent the results obtained form equation (2.8) using the calculated values of dielectric properties.

Figure 2.16 shows the results of the measured *SWR* in (dB) for the four samples with different levels of porosity as a function of sample thicknesses at 12 GHz. This figure clearly demonstrates the importance of making measurements using many sample thicknesses when utilizing this measurement approach. For example if the thickness of the samples are chosen to be around 47.5 mm, there would be no distinction among porosity levels of 0%, 48.8% and 58.7%. However, at a thickness of around 43 mm there is clear difference among three of the samples. Clearly, using many sample thicknesses in the measurements and the subsequent calculations aids in distinguishing among these porosity levels.

Tables 2.8 and 2.9 show the results of the relative permittivity and loss factor for these samples. The results show little variation of ε'_r and ε''_r as a function of frequency, which is expected since these samples have low relative permittivities and loss factors. Any change in these values, as a function of frequency, is considered within the measurement uncertainty. However, there is a clear distinction among different air volume contents. ε'_r and ε''_r for the 0% porosity samples indicate the dielectric properties of the basic sample (i.e. not porous). The difference in porosity level of about 10% among the other three samples causes a nearly linear change in the values of ε'_r and ε''_r. However, one must be careful not to generalize this trend for composites with different dielectric properties than those used here. For example, in ceramics with higher relative permittivities (ε'_r = 6-20) one would expect a larger change in the dielectric properties for linear changes in porosity levels.

Using the measurement results at 10 GHz, one percent change in porosity level translates to about 1.1% change in ε'_r and 3% change in ε''_r, respectively. This means that if the relative permittivity and the loss factor can be measured to within these percentages, then it is possible to detect 1% change in porosity for this dielectric material. As already mentioned, there are several non-destructive microwave techniques for dielectric properties monitoring, and each technique has its own specific range of measurement accuracy (yielding different sensitivity to porosity variation). The same measurement accuracy analysis, as that explained in Section 2.3.5, was conducted for these measurements at 10 GHz and 16 GHz. The results for these two frequencies showed the measurement uncertainty to be about

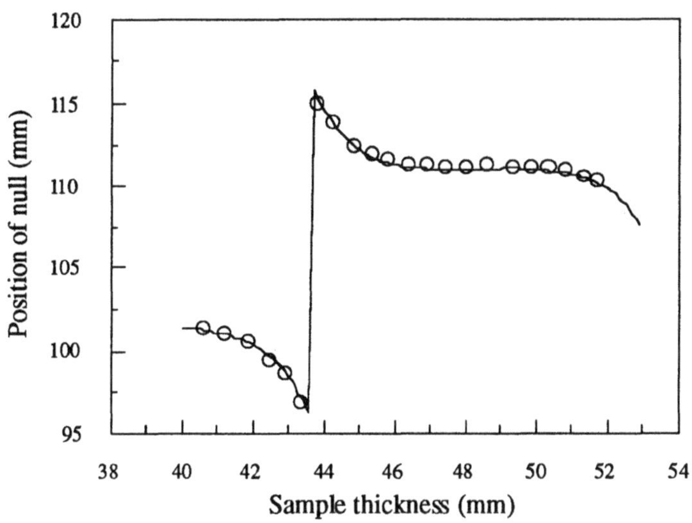

Fig. 2.14 Position of null vs. sample thickness at 10 GHz for 0% porosity (Gray *et al.*, 1995).

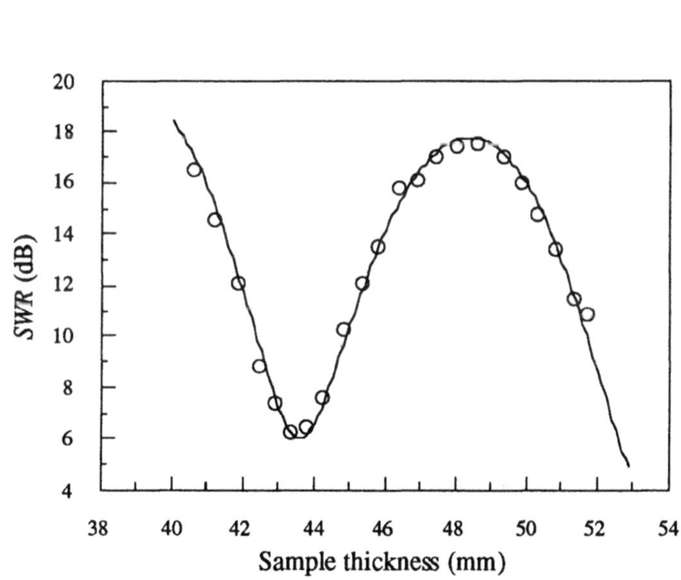

Fig. 2.15 The *SWR* (dB) vs. sample thickness at 10 GHz for 0% porosity (Gray *et al.*, 1995).

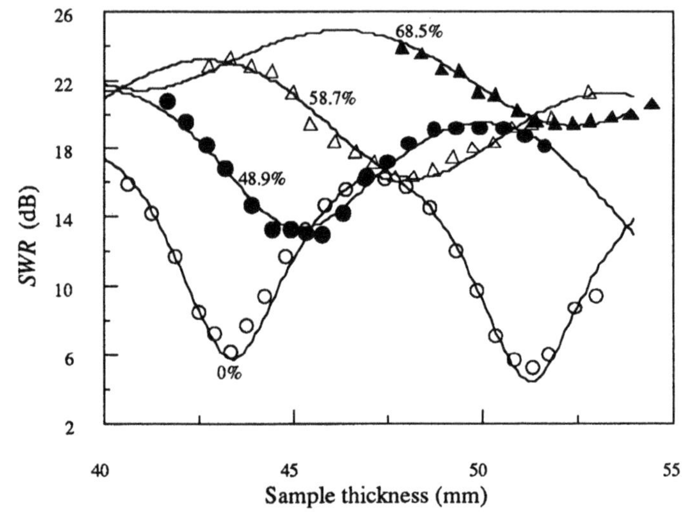

Fig. 2.16 The *SWR* (dB) vs. sample thickness at 12 GHz for four different porous samples (Gray *et al.*, 1995).

Table 2.8 ε'_r for 0%, 48.9%, 58.7% and 68.5% air content in polymer microballoon-filled epoxy (Gray *et al.*, 1995).

Frequency (GHz)	*Air content*			
	0%	*48.9%*	*58.7%*	*68.5%*
8.2	2.80	1.87	1.69	1.48
10	2.87	1.84	1.63	1.46
12	2.83	1.88	1.70	1.47
14	2.87	1.83	1.70	1.50
16	2.84	1.82	1.68	1.47
18	2.84	1.84	1.67	1.47

Table 2.9 ε''_r for 0%, 48.9%, 58.7% and 68.5% air content in polymer microballoon-filled epoxy (Gray *et al.*, 1995).

Frequency (GHz)	*Air content*			
	0%	*48.9%*	*58.7%*	*68.5%*
8.2	0.086	0.032	0.020	0.013
10	0.086	0.034	0.023	0.015
12	0.082	0.033	0.022	0.014
14	0.083	0.033	0.027	0.022
16	0.077	0.026	0.021	0.019
18	0.068	0.027	0.025	0.014

1.2% for the relative permittivity about than 5% for the loss factor. Thus, using the change in the dielectric properties as an indicator of porosity level change, it is possible to determine porosity changes of slightly more than 1% (when using ε'_r) and about 2% (when using ε''_r). Since there is little variation in ε'_r and ε''_r as a function of frequency, the same measurement accuracy is assumed for the other frequencies.

2.6 RUST DIELECTRIC PROPERTIES[4]

Metals, specially steels, are susceptible to corrosion when in moist and salty environments. When painted, the corrosion or rust is not always visually detected unless it has become relatively severe and the paint begins to blister (Funke, 1981; Collins, 1993). Naval ships, commercial shipping vessels, steel bridges and steel platforms in oceans are examples of some of the environments in which severe corrosion may occur. Consequently, detection of corrosion under paint and other dielectric coatings is an important practical concern. Near-field microwave non-destructive testing techniques, employing rectangular waveguide probes, have shown great promise in detecting and evaluating rust under such coatings (Qaddoumi, Shroyer and Zoughi, 1997). However, in order to theoretically (and subsequently experimentally) investigate the potential of such techniques, the dielectric properties of paint, primer, composite laminates of interest and rust must be measured. Therefore, in the original investigation chemically produced iron oxide powder, Fe_2O_3, was used to represent rust in the theoretical simulations (Qaddoumi, Shroyer and Zoughi, 1997). Subsequently, the dielectric properties of Fe_2O_3 were measured. Fe_2O_3 was found to be a low-permittivity and low-loss dielectric material, consistent with the dielectric properties of other metal oxides (see Table 2.1). However, real rust is expected to have considerably different dielectric properties than Fe_2O_3 powder. This is due to the fact that some iron may be present in real rust which tends to increase its permittivity as well as its loss factor. Moreover, the addition of salts may also influence the dielectric properties of real rust. Consequently, several different rust specimens were obtained and their dielectric properties were measured.

---

[4] Portions of this section are reprinted with permission from *Materials Evaluation* (in press), Copyright © The American Society for Nondestructive Testing, Inc.

2.6.1 Measurement procedure

The dielectric properties of various rust specimens were measured using a completely-filled two-port transmission line techniques, as shown in Fig. 2.17. In this approach a portion of a straight section of a rectangular waveguide (sample holder) is filled with a rust specimen which is compacted to fill the entirely of the waveguide sample holder. The straight waveguide section is terminated on both ends with two waveguide-to-coax adapters. Subsequently, a vector network analyzer (in this case an HP 8510) is used to measure the scattering parameters of this two-port transmission line system (Pozar, 1998). The scattering parameters are related to the reflection coefficient through (S_{11}) and transmission coefficient through (S_{21}), as shown in Fig. 2.17. These two parameters are measured by the network analyzer in the waveguide frequency band (or at a single frequency). These parameters are then used to calculate the relative permittivity and loss factor of the dielectric specimen filling the sample holder (e.g. rust powder). The foundations of this technique is described in detail in (Baker-Jarvis, 1990; Nicholson and Ross, 1970; Weir, 1970). Later this technique was expanded to accommodate dielectric property measurement of liquids by adding two dielectric plugs around dielectric sample holder (Bois, 1999; Handjojo, 1999). This improved technique is designed to alleviate the meniscus problem associated with short-circuited waveguide measurements (as discussed for resin binder in Section 2.4).

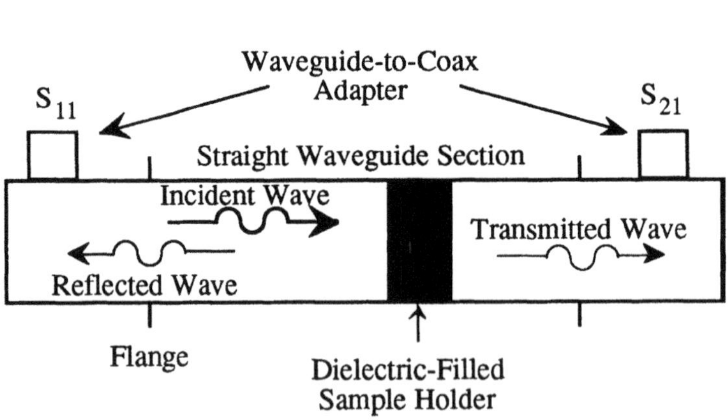

Fig. 2.17 The schematic of a completely-filled two-port transmission line dielectric property measurement technique.

2.6.2 Rust specimen descriptions and measured dielectric properties

Using the measurement procedure described above, the dielectric properties of typical paint and primer as well as several types of rust specimens were measured (Qaddoumi, Handjojo, Bigelow *et al.*, 1999). The following is the description of each specimen:

- *Red Rust* - This fine rust powder specimen was obtained from steel plates that had been rusted in a hydration room. This is considered to be one of the most common types of rust.
- *Black Rust* - This rust specimen was obtained by crushing, into a powder, solid granules of rusted steel in which the steel may not have completely turned into rust. This specimen was used for comparison with Fe_2O_3 powder and red rust.
- *Hydrated Black Rust* - The black rust mentioned above was left in a hydration room for five days to see whether more of the steel would turn into rust.
- *Salt Rust* - This specimen was obtained from steel in left in salt water containing approximately 37% salt to simulate rusting in steel in salt water.
- *Fe_2O_3 powder* - Fe_2O_3 powder was obtained and its dielectric properties had already been measured (Qaddoumi, Shroyer and Zoughi, 1997).
- *Paint* - Typical paint was sprayed into specially prepared Teflon forms, and dried into several millimeter-thick sheets (after they were peeled off).
- *Primer* - Typical primer was also sprayed on paint samples and dried into several millimeter-thick sheets. The results were almost identical to those of paint, as expected.

Table 2.10 shows the measured average values of ε'_r and ε''_r for various rust specimens, paint and primer at X-band

Table 2.10 Measured average values of ε'_r and ε''_r for various rust specimens, paint and primer at X-band (Qaddoumi, Handjojo, Bigelow *et al.*, 1999).

Specimen	ε'_r	ε''_r
Red rust	8.42	1.03
Black rust	12.58	2.36
Hydrated black rust	11.28	2.05
Salt rust	5.33	0.53
Paint-primer	3.48	0.12
Fe_2O_3 powder	3.22	0.04

The results indicate that the dielectric properties of paint and primer are in the family of low-permittivity and low-loss materials. The same is true for Fe_2O_3 powder. However, red rust has a relative permittivity which is about 2.5 times larger than that of paint. More importantly, the loss factor of red rust is greater than that of paint by about an order of magnitude. Black rust has a relative permittivity of about four times larger than that of paint, and a loss factor of around twenty times more than that of paint. The difference in the dielectric properties of rust and paint (and primer) coupled with the sensitive technique of using an open-ended rectangular waveguide probe for inspecting layered dielectric structures, renders the presence of slight amount of rust under paint easily detectable (Qaddoumi, Shroyer and Zoughi, 1997; Qaddoumi, Handjojo, Bigelow *et al.*, 1999).

2.7 DIELECTRIC MIXING MODELS[5]

The average dielectric properties of a dielectric composite material or a mixture composed of several different constituents (e.g. carbon black loaded rubber) can be related to:

- the dielectric properties of each of its constituents,
- each of their volume contents (or fractions),
- their spatial distribution within the mixture (homogeneity),
- the relative orientation of constituents with respect to the impressed electric field (anisotropy),
- the influence of any chemical reactions such as curing.

In addition to curing, other molecular reactions can significantly influence the dielectric properties of a mixture. A good example of this is the presence of water in a mixture. Depending on whether the water molecules are free or bound, the influence of water dielectric properties (Bois, Benally, Nowak *et al.*, 1998; Ulaby, Moore and Fung, 1986), on the average dielectric properties of a mixture, can be dramatically different. For today's complex composite materials some of the factors mentioned above may be more influential in determining their average dielectric properties than others. As was discussed in Section 2.3.4, a volumetrically small addition of curatives to the rubber mixture caused significant change in the dielectric properties of rubber compounds (i.e. molecular linking influence). Similarly, the presence of carbon black which is a high-

[5] Portions of this section are reprinted with permission from *Materials Evaluation*, vol. 53, no. 3, pp. 404-08, Copyright 1995 © The American Society for Nondestructive Testing, Inc.

permittivity and high-loss material dominated the dielectric properties of cured rubber (i.e. dielectric property and volume fraction influence).

Dielectric mixing models primarily take into account the five factors mentioned above to predict the average dielectric properties of a mixture. Such mixing models generally have an electromagnetic basis (Hasted, 1973; Taylor 1966; Korneeko, 1960). However, for many mixtures it may not be possible to account for all electromagnetic influences and interactions among the various constituents. Therefore, many approximations have been introduced in ways by which the average dielectric properties of a mixture is related to its constituents according to the five factors mentioned earlier. These approximations include the assumption of uniformity of the size of constituents and uniformity in their spatial distribution, etc. (Tinga, Voss and Blossey, 1973; de Loor, 1964; de Loor, 1968). For a comprehensive review of dielectric mixing models one may refer to the following references (Ulaby, Moore and Fung, 1986; Van Beek, 1967; Bottcher, 1952). A dielectric mixture is considered to be composed of a host medium (the most prominent constituent) and several types of inclusions. Appendix E of (Ulaby, Moore and Fung, 1986) provides for a fairly detailed discussion of these models as they relate to earth matters such as water, soils and snow.

Similar approaches may be used to develop models for any dielectric composite. However, due to the complexity of these materials and a lack of knowledge regarding the chemical, molecular and microscopic interaction of various constituents in today's composite materials, one may develop what is known as an empirical dielectric mixing model. In such models the composite is assumed to be made of a host medium and several inclusions. Subsequently, the knowledge of the host and each constituent dielectric properties and their respective volume fractions are used to obtain an empirical expression that predicts the average dielectric properties of the mixture. These empirical expressions, their validity and usefulness are based on the manner by which their empirical constants are obtained. These constants are usually found by making numerous dielectric property measurements of the composite material as a function of varying constituent properties and volume contents. Subsequently, through fitting the empirical expression to the results of measurements, the unknown empirical constants are obtained. In the next section two examples of these empirical dielectric mixing models are presented.

2.7.1 Empirical dielectric mixing model for cured carbon black loaded rubber

It is intended to develop a dielectric mixing model for cured carbon black loaded rubber based on the measured results depicted in Figs. 2.5 and 2.6 without considering the influence of any chemical reaction. The goal is to show that a simple empirical dielectric mixing model can be developed to predict carbon black percentage in properly cured rubber. Clearly, for accurate prediction one must eventually bring into this model the influence of the curatives as well. As explained earlier, there are various types of dielectric mixing models available. A simple two-phase mixing model is considered for this case which accounts for the dielectric properties of the mixture, in this case the cured rubber, as a combination of a host medium (EPDM) and an inclusion (carbon black). Multi-phase dielectric mixing models allow for taking into account more than one type of inclusion in the host medium (Ulaby, Moore and Fung, 1986). These models are usually for heterogeneous media in which the dielectric properties of the constituents are not very different. In the present case carbon black has very high conductivity, and hence high dielectric properties. Knowing that the dielectric properties of all other rubber constituents are similar (see Table 2.1), the dielectric properties of rubber may be modeled as a two-phase dielectric medium with EPDM as the host medium and carbon black as the inclusion.

Consequently, an empirical dielectric mixing expression, suited for cured carbon black loaded rubber, was developed. The approach described in (Ulaby, Moore and Fung, 1986) which pertains to soil moisture monitoring was used as the basis for this model. Equation (2.12) expresses the relationship between rubber compound mixture dielectric property, ε_m, dielectric properties of EPDM (host), ε_h, and carbon black (inclusion), ε_c, and their volume fractions.

$$\varepsilon_m(\alpha,\beta,\gamma,V)=\gamma\left[V^{\beta}\,\varepsilon_c^{\alpha}+(1-V_c)\varepsilon_h^{\alpha}\right]^{\frac{1}{\alpha}} \tag{2.12}$$

where V_c is the volume fraction of carbon black and α, β, γ are unknown coefficients. The unknown coefficients are calculated by setting equation (2.12) equal to the measured value of dielectric properties of rubber compound with different carbon black contents. Figures 2.18 and 2.19 show the results of equation (2.12) and the measured relative permittivity and loss factor of rubber compound for all carbon black volume fractions at 5 GHz. The good agreement between the model and the measurement results indicates that the properties of the cured rubber may be predicted using a relatively simple mixing dielectric model. The unknown coefficients are generally frequency dependent, and must be re-evaluated for another frequency.

2.7.2 Empirical dielectric mixing model for microballoon-filled epoxy resin

Since the polymer microballoons and the epoxy resin, as described in section 2.5, have very similar dielectric properties, these samples were considered to be made of only two constituents; namely, air and polymer, and requiring only a two-phase dielectric mixing model. The air inclusions are in the form of microballoons, hence they are considered to be spherical in shape. For these reasons, the following well known two-phase spherical mixing model was used to predict the dielectric properties as a function of porosity level or air volume fraction (Ulaby, Moore and Fung, 1986).

$$\varepsilon_m = \varepsilon_h + 3V_i\varepsilon_m\left(\frac{\varepsilon_i - \varepsilon_h}{\varepsilon_i + 2\varepsilon_m}\right) \qquad (2.13)$$

where ε_m, ε_i and ε_h are the dielectric properties of the composite mixture, inclusion (air), and the host medium (e.g. the dielectric properties of the sample with 0% porosity), respectively, and V_i is the volume fraction of the air inclusion (porosity).

Figures 2.20 and 2.21 show the relative permittivity and loss factor obtained from the model (solid line), along with the measured values (circles) at a frequency of 10 GHz. The agreement between the results of this two-phase dielectric mixing model and the measurements is good. Thus, it is possible to either predict the dielectric properties of this composite as a function of porosity level, or more importantly by measuring the dielectric properties of a sample one is able to determine the level of porosity in it (e.g. V_i).

2.8 SUMMARY

Material characterization is an extremely important issue in various industries. The objective of this chapter has been to introduce the reader to the issue of material characterization using dielectric properties of materials. Similar approaches may be undertaken to evaluate the characteristics of other materials in much the same manner.

The dielectric property of materials as an important measurement parameter was briefly introduced from a basic point of view. The effect of alternating fields on this parameter was also discussed and the issue of

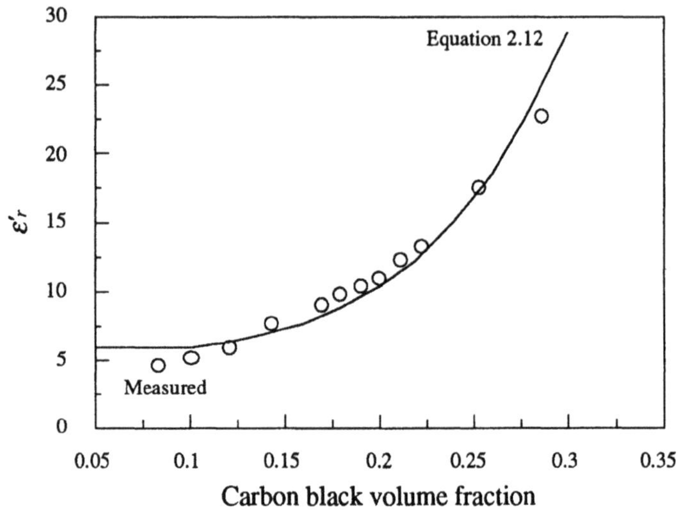

Fig. 2.18 Relative permittivity for cured rubber vs. carbon black volume fraction at 5 GHz measured and calculated using the mixing model.

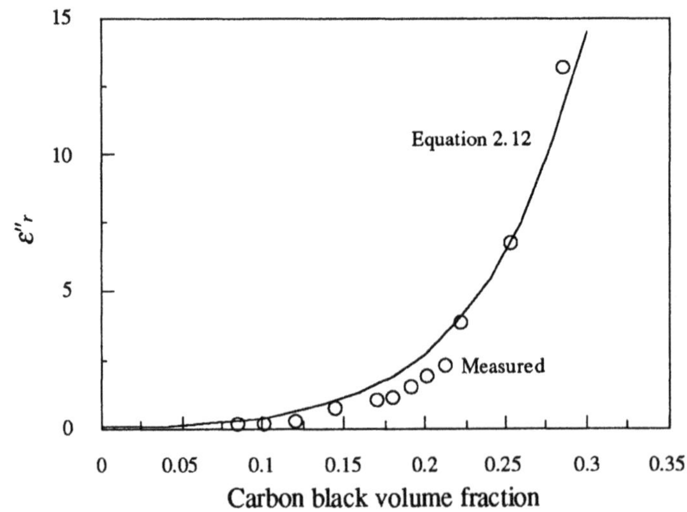

Fig. 2.19 Relative loss factor for cured rubber vs. carbon black volume fraction at 5 GHz measured and calculated using the mixing model.

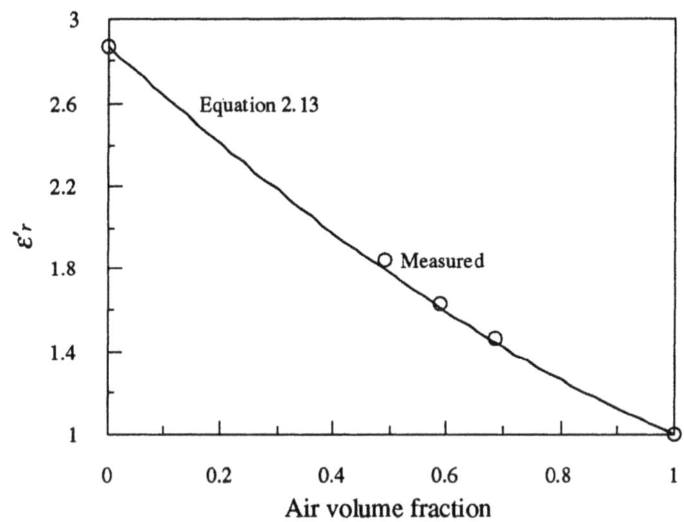

Fig. 2.20 Relative permittivity for microballoon-filled polymer epoxy vs. air volume fraction at 10 GHz measured and calculated using the mixing model (Gray *et al.*, 1995).

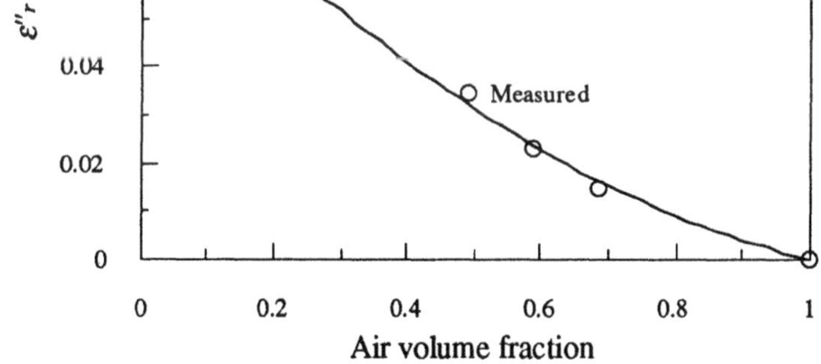

Fig. 2.21 Relative loss factor for microballoon-filled polymer epoxy vs. air volume fraction at 10 GHz measured and calculated using the mixing model (Gray *et al.*, 1995).

energy storage and energy absorption by generally lossy dielectric materials were discussed. Using the well known completely-filled waveguide approach the dielectric properties of several mixtures were evaluated.

In properly cured carbon black loaded rubber, it was shown that carbon black volume content can be closely estimated. In uncured rubber, it was found that the presence of curatives can be detected even when the rubber has not been properly cured. In other words, the initiation of curing at room temperature in these samples was detected. This is an important finding since in rubber industry many millions of dollars are lost as a result of not including curatives in rubber compounds prior to the curing process. It was also shown that curing is a complex phenomenon which includes the interaction of EPDM, curatives and carbon black.

Investigations on liquid resin binder showed that a slight amount of curing, as a result of leaving resin binder at room temperature for 12 days, can be detected. Low density fiberglass materials used for insulation purposes incorporate resin binders in their manufacturing process. Subsequently, the binder must be properly cured before the product is ready for use. In these investigations it was shown that the loss factor associated with five different fiberglass samples may be used for determining their level of resin binder and their state of cure. However, the measurement confidence associated with these measurements may not be high enough. As will be shown in Chapter 5, an open-ended rectangular waveguide probe is found to provide a much higher measurement confidence when distinguishing among these samples.

Polymer microballoon-filled epoxy resin was used to determine whether dielectric property characterization can be utilized to estimate porosity level in such materials. Different porosity levels were achieved using different volume fractions of microballoons. This investigation showed that porosity level variations of about 1% can be determined in these materials. These encouraging results may now open the door for investigating porosity variation in other plastics, ceramics and composite mixtures in a similar fashion.

Finally, dielectric mixing models were introduced. The utility of simple empirical dielectric mixing models as a tool to predict the average dielectric property of a mixture was shown. Conversely, these models may be used to determine the volume fraction or the dielectric properties of a certain constituent in a mixture. Two simple expressions were introduced to predict the average dielectric properties of cured rubber compounds as a function of carbon black volume content and polymer microballoon-filled epoxy as a function of air content.

REFERENCES

Altschuler, H.M. (1963) *Dielectric Constant*, ch. 9 in Handbook of Microwave Measurements, vol. II, (eds. M. Sucher and J. Fox), Polytechnic Press, Brooklyn, New York.

Armistead, R. A. and Yancey, R.N. (1989) Computed tomography for the nondestructive testing of advanced engineering materials. *Materials Evaluation*, **47** (5), pp. 487-91.

Baker-Jarvis, J. (1990) Transmission/reflection and short-circuit line permittivity measurements, NIST Technical Note 1341, U.S. Government Printing Office, Washington, D.C., July.

Baker-Jarvis, J., Jones, C., Riddle, B. *et al.* (1995) Dielectric and magnetic measurements: A survey of nondestructive, quasi-nondestructive, and process-control techniques. *Research in Nondestructive Evaluation*, **7** (2/3), pp. 117-36.

Bakhtiari, S., Ganchev, S.I. and Zoughi, R. (1993) Microwave frequency optimization for accurate thickness or dielectric property monitoring of conductor backed composites. *Materials Evaluation*, **51** (6), pp. 740-43.

Bakhtiari, S., Ganchev, S.I., Qaddoumi, N. *et al.* (1994) Microwave non-contact examination of disbond and thickness variation in stratified composite media. *IEEE Transactions on Microwave Theory and Techniques*, **42** , pp. 389-95.

Balanis, C.A. (1989) *Advanced Engineering Electromagnetics*, John Wiley & Sons, New York.

Bois, K.J. (1999) Near-field microwave inspection and characterization of cement based materials, Ph.D. Dissertation, Electrical and Computer Engineering Department, Colorado State University, January.

Bois, K.J., Benally, A.D., Nowak, P.S. *et al.* (1998) Cure-state monitoring and water-to-cement ratio determination of fresh Portland cement based materials using near-field microwave techniques, *IEEE Transactions on Instrumentation and Measurement*, **47** (3), pp. 628-637.

Bottcher, C.J.F. (1952) *Theory of Electric Polarization*, Elsevier, Amsterdam.

Bussey, H.E. (1967) Measurement of RF properties of materials. A survey. *Proceedings of IEEE*, **55**, pp. 1046-53.

Chao, S. H. (1986) An uncertainty analysis for the measurements of microwave conductivity and dielectric constant by the short-circuited line method. *IEEE Transactions on Instrumentation and Measurement*, **35**, 36-41.

Collins, J.A. (1993) *Failure of Materials in Mechanical Design*, 2nd Ed., Wiley Intersciene, New York.

Daniel, M., Wooh, S.C. and Komsky, I. (1992) Quantitative porosity characterization of composite materials by means of ultrasonic attenuation measurements. *Journal of Nondestructive Evaluation*, **11** (1), pp. 1-8.

de Loor, G.P. (1964) Dielectric properties of heterogeneous mixtures. *Applied Scientific Research*, **B11**, pp. 310-20.

de Loor, G.P. (1968) Dielectric properties of heterogeneous mixtures containing water. *Journal of Microwave Power*, **3**, pp. 67-73.

Funke, W. (1981) Blistering of Paint Films, in *Corrosion by Organic Coatings*, (ed. H. Leidheiser, Jr.), National Association of Corrosion Engineers, Houston, TX, pp. 97-102.

Ganchev, S., Qaddoumi, N., Brandenburg, D. *et al.* (1994) Microwave diagnosis of rubber compounds. *IEEE Transactions on Microwave Theory and Techniques*, **42** (1), pp. 18-24.

Gray, S, Ganchev, S.I., Qaddoumi, N. *et al.* (1995) Porosity level estimation in polymer composites using microwave. *Materials Evaluation*, **53** (3), pp. 404-08.

Handjojo, L. (1999) Sensitivity analysis of two-port transmission line technique for dielectric property measurement using dielectric plugs, M.S. Thesis, Electrical and Computer Engineering Department, Colorado State University, February.

Hashimoto, O. and Shimizu, Y. (1986) Reflecting characteristic of anisotropic rubber sheets and measurement of complex permittivity tensor. *IEEE Transactions on Microwave Theory and Techniques*, **34**, 1202-07.

Hasted, J.B. (1972) Liquid water: Dielectric properties, ch. 7 in *Water-A Comprehensive Treatise*, (ed. F. Franks) **1**, Plenum Press, New York.

Jones, T. and Berger, H. (1989) Application of nondestructive inspection methods to composites. *Materials Evaluation*, **47** (4), pp. 390-400.

Jow, J., Hawley, M.C., Finzel, M. *et al.*, (1987) Microwave processing and diagnosis of chemically reacting materials in a single-mode cavity applicator. *IEEE Transactions on Microwave Theory and Techniques*, **35**, 1435-43.

Korneenko, I.A. (1960) Mean values of the parameters in inhomogeneous media. *Soviet Physics-Technical Physics*, **5**, p. 40.

Kraus, J.D. (1992) *Electromagnetics*, 4th Ed., McGraw Hill, New York, p. 804.

Lipowitz, J., Rabe, J. A., Frevel, L.K. *et* al. (1990) Characterization of nanoporosity in polymer derived ceramic fibers by X-ray scattering techniques. *Journal of Materials Science*, **25**, pp. 2118-24.

Martinelli, M., Rolla, P.A. and Tombari, E. (1985) A Method for dynamic dielectric measurements at microwave frequencies: applications to polymerization process studies. *IEEE Transactions on Instrumentation and Measurement*, **34**, 417-21.

Nicholson, A.M. and Ross, G. (1970) Measurement of intrinsic properties of materials by time domain techniques, *IEEE Transactions on Instrumentation and Measurement*, **19** (4), pp. 377-382.

Pozar, D.M. (1998) *Microwave Engineering*, Second edition, John Wiley and Sons, New York.

Qaddoumi, N., Ganchev, S.I. and Zoughi, R. (1996) Microwave diagnosis of low-density fiberglass composites with resin binder. *Research in Nondestructive Evaluation*, **8** (3), pp. 177-88.

Qaddoumi, N., Handjojo, L., Bigelow, T. *et al.* (in press) Microwave corrosion detection using open-ended rectangular waveguide sensors, *Materials Evaluation.*

Qaddoumi, N., Shroyer, A. and Zoughi, R. (1997) Microwave Detection of Corrosion Under Paint and Composite Laminate Coatings. *Research in Nondestructive Evaluation*, **9** (4), pp. 201-212.

Ramo, S., Whinnery, J.R. and Van Duzer, T. (1994) *Fields and Waves in Communication*, 3rd Ed., Wiley, New York.

Roberts, S. and von Hippel, A. (1946) A new method for measuring dielectric constants and loss in the range of centimeter waves. *Journal of Applied Physics*, **17**, pp. 610-16.

Rose, R.M., Shepard, L.A. and Wulff, J. (1966) *Electronic Properties*, Wiley, New York, pp. 262.

Rubin, A. M. and Jerina, K.L. (1993) Evaluation of porosity in composite aircraft structures. *Composites Engineering*, **3** (7-8), pp. 601-18.

Sardos, R., Escarmant, J.F. and Saint-Christophe, E. (1990) Anisotropic measurements of rubber sheets with and X-band three-wave interferometer. *IEEE Transactions on Microwave Theory and Techniques*, **38**, 330-33.

Schwarz, H.F., Bosisio, R.G. and Wertheimer, M.R. *et al.* (1973) Microwave curing of synthetic rubbers. *Journal of Microwave Power*, **8**, 303-22.

Smith, D. M., Hua, D.W. and Earl , W.L. (1994) Characterization of porous solids. *Materials Research Society Bulletin*, **19** (4), pp. 44-48. Taylor, L.S. (1965) Dielectric properties of mixtures. *IEEE Transactions on Antennas and Propagation*, **13**, pp. 943-47.

Taylor, L.S. (1966) Dielectric loaded with anisotropic materials. *IEEE Transactions on Antennas and Propagation*, **14**, pp. 669-70.

Terselius, B. and Ranby, B. (1978) Cavity perturbation measurements of the dielectric properties of vulcanizing rubber and polyethylene compounds. *Journal of Microwave Power*, **13**, 327-35.

Thompson, R. B., Thompson, D.O., Holger, D.K. *et al.* (1991) Enhancing analysis techniques for composite materials. *Ultrasonic NDE of Thick Composites*, ASME , **10**, pp. 43-57.

Tinga, W.R., Voss, W.A.G. and Blossey, D.F. (1973) Generalized approach to multiphase dielectric mixture theory. Journal of *Applied Physics*, **44**, pp. 3897-902.

Tsankov, M.A. (1981) Measurable values of permittivity and loss tangent of dielectrics and ferrites for waveguide microwave methods. *Bulgarian Journal of Physics*, **8** (4), pp. 403-14.

Ulaby, F.T., Moore, R.K. and Fung, A.K. (1986) *Microwave Remote Sensing, Active and Passive*, Vol. 3, Appendix E, Artech House, Dedham, MA.

van Beek, L.K.H. (1967) Dielectric behavior of heterogeneous systems, ch. 7 in *Progress in Dielectrics*, (ed. J. Birks), Heywood Books, London.

von Hippel, A.R. (1954) *Dielectrics and Waves*, MIT Press, Cambridge, MA.

Weir, W.B. (1970) Automatic measurement of complex dielectric constant and permittivity at microwave frequencies, *Proceedings of IEEE*, **62** (11), pp. 33-36.

Wright, R.R. and Skutt, K.H.R. (1965) *Electronics: Circuits and Devices*, Ronald Press, New York.

CHAPTER 3

Layered dielectric composite evaluation

3.1 INTRODUCTION

Dielectric sheet materials are extensively used in a variety of applications. Furthermore, composites made of several such layers, adhesively bonded together, are finding their way as structural members in a broad range of applications as well. Whether one is interested in obtaining accurate information about the thickness of a rubber sheet or a thin plastic film or inspecting a thick multi-layer (sandwich) composite structure for the presence of disbonds and delaminations, the inspection technique must be capable, robust and easy to use. When inspecting such composites, standard non-destructive techniques may not always be suitable particularly when these composites are relatively thick and are made of porous, low-permittivity and low-loss dielectric materials. In addition, the usual planar geometry of defects, such as disbonds, limit the application of some of the standard non-destructive techniques. Ultrasonic techniques, which usually require a couplant, are of limited use due to the high attenuation experienced by the signal when inspecting thick and porous composites. X-ray radiography may also be used for inspection of thick composites, however it has some limitations. Some of these limitations may be overcome using neutron radiography, but equipment size and weight make this approach limited in its practicality (Bovig, 1989; Carriveau, 1994). Microwave signals can penetrate deep inside low-loss dielectric materials and interact with their inner structure without suffering from high attenuation. They are also sensitive to changes associated with boundary interfaces, which makes them very attractive for disbond detection in sandwich composite structures (Bahr, 1982; Zoughi and Bakhtiari, 1990). Microwave approaches do not require a couplant, and may be applied either in contact or non-contact fashion. When used in a non-contact fashion and in the near-field of a probe, the standoff distance may be used as an optimization parameter to achieve increased sensitivity to the presence of defects such as disbonds.

Microwave non-destructive testing techniques have shown great potential for determining dielectric sheet thicknesses in stratified media along with accurate dielectric sheet thickness variation and disbond detection (Zoughi and Bakhtiari, 1990; Bakhtiari *et al.*, 1994). These methods vary significantly in their basic approaches. Consequently, the measurement sensitivity achieved by each method is also different. Some of these approaches have also been extended to measurement-parameter optimization which has resulted in significant improvement in their measurement sensitivity (Zoughi and Lujan, 1990; Zoughi and Bakhtiari, 1990; Bakhtiari and Zoughi, 1990; Bakhtiari *et al.*, 1994, Ganchev *et al.*, 1995a; Bakhtiari, Gopalsami and Raptis, 1995; Han and Mansueto, 1995; Han and Mansueto, 1997; Ganchev, *et al.*, 1995b; Ganchev, *et al.*, 1995c; Bakhtiari, Ganchev and Zoughi, 1993a; Bakhtiari, Ganchev and Zoughi, 1993b; Bakhtiari, Ganchev and Zoughi, 1995; Gray and Zoughi, 1997). Microwave near-field techniques employing open-ended waveguide sensors provide high level of measurement sensitivity to sheet thickness variation and disbond detection. When using such a microwave sensor two measurement parameters, the standoff distance and the operating frequency may be varied or optimized to achieve optimal sensitivity to thickness variation and disbond detection. Measurement-parameter optimization may be first numerically achieved using an electromagnetic model (Bakhtiari *et al.*, 1994; Bakhtiari, Ganchev and Zoughi, 1995; Ganchev *et al.*, 1995b), and subsequently applied experimentally. Once a parameter of interest is chosen (i.e. thickness variation of a dielectric sheet), such models can be used to investigate the influence of variations in all other measurement parameters (i.e. standoff distance, operating frequency, etc.).

3.2 WHAT MAY BE ACCOMPLISHED

Microwave non-destructive testing and evaluation techniques may be applied to many aspects of composite processing and manufacturing. In Chapter 2, the utility of microwave techniques for material characterization and cure-state monitoring was illustrated. Other application to which microwave techniques may be applied include process control when manufacturing single and multi-layered dielectric composites. These may include ensuring the production of a dielectric sheet within a given thickness tolerance, producing layered composites in which the integrity of the bonded layers must be ensured, detecting delamination in thick dielectric structures, etc. When concerned with manufacturing and process control of composite structures the following may be accomplished using microwave non-destructive techniques:

- Absolute thickness of dielectric sheets may be accurately determined.
- Minute thickness variation in dielectric sheets may be evaluated.
- These dielectric sheets may be backed by a conducting plate, free-space, an infinite half-space of another dielectric material or be part of a layered dielectric structure.
- Very thin disbonds may be identified.
- Slight delaminations in a dielectric may be identified.
- Relatively fine spatial resolution may be obtained.
- Given the thickness of a dielectric sheet backed by free-space or a conducting plate its dielectric properties may be accurately determined.
- All of the above may be conducted in a contact or non-contact fashion.
- All of the above may be performed in real-time and on an on-line basis.
- In all of the above applications access to only one side of the material is needed (i.e. reflection measurements).

There are several approaches by which the above tasks may be accomplished. However, the two prominent techniques include the far-field (i.e. radar techniques) and near-field techniques (focus of recent intense research). In the former technique the specimen under inspection is in the far-field of a sensor which is usually some type of a microwave antenna. In the latter technique the specimen under test is placed near the microwave sensor. Each approach has its advantages and disadvantages. However, the near-field approaches are more suited for the overwhelming majority of non-destructive applications mentioned above. A brief description of various field regions as well as a description of the attributes of each approach is presented next.

3.3 FIELD REGIONS

Commonly, there are three regions around a microwave source, antenna or sensor (Balanis, 1997). The region immediately surrounding the source is known as the reactive near-field in which reactive (stored) fields are predominant. In this region quasi-static fields exist. Open-ended coaxial sensors commonly operate in this region. Next, is the radiating near-field region. In this region reactive and radiating fields exist, however, the spatial distribution of the radiating fields varies as a function of distance to the source. Open-ended rectangular waveguides, for the purpose of applications mentioned in the previous section, operate in this region. Finally, the far-field region is defined as the region in which primarily radiating fields exist. The spatial distribution of the fields in this region is independent of the distance from the source. Plane waves or TEM (transverse electromagnetic) waves are assumed to exist in this region.

3.3.1 Far-field approach

For the purpose of inspecting layered dielectric structures, far-field approaches have been used to determine dielectric sheet thicknesses (Zoughi and Lujan, 1990; Ulaby, Moore and Fung, 1981), presence of voids and disbonds (Zoughi and Bakhtiari, 1990; Bakhtiari, Gopalsami and Raptis, 1995) and dielectric properties of sheets (Zoughi and Zonnefeld, 1991). Generally, a relatively large footprint, illuminating the specimen under inspection, is associated with this technique. This causes problems due to edge reflections around specimens with small area extents. Furthermore, in most cases where the phase of the reflection or transmission coefficient is sought, calibration of the plane of reference, which is usually the front surface of the sheet, is required. For most practical applications this may not be an easy task to accomplish. Finally, the results of this approach, either thickness measurement, disbond detection or dielectric property characterization are not very accurate, and do not provide as fine of a resolution as those obtained using near-field techniques. Far-field techniques usually require a microwave antenna. The theoretical electromagnetic modeling of plane waves interacting with layered dielectric structures is relatively simple for which transmission line or field intensity modeling approaches may be utilized (Zoughi and Bakhtiari, 1990; Ulaby, Moore and Fung, 1981). For a thorough derivation of the interaction of plane waves with layered dielectric structures the reader my consult the above two references.

3.3.2 Near-field approach

Near-field techniques are most suitable for the majority of practical non-destructive evaluation applications, in particular when inspecting thick sandwich composites. Some of the reasons for this are:

- The spatial resolution, when operating in this region, is primarily influenced by the scanning probe dimensions rather than the operating wavelength. Therefore, compared to the plane-wave approach much finer spatial resolutions may be obtained.
- Contact as well as non-contact type of measurements are possible.
- Having access to almost all areas of a composite structure, with complicated geometry, is possible without much consideration given to unwanted reflection such as edge effects.

- Calibration, when seeking thickness variation, disbond detection, delamination detection and even absolute thickness evaluation, is either not necessary or it is a relatively simple task.
- There are many open-ended probes that may be used; such are open-ended rectangular waveguides, open-ended circular waveguides, open-ended waveguides of other aperture geometry, open-ended coaxial lines, open cavity resonators, microstrip patches, small horn antennas as long as one is operating in their near-field regions, etc. Each of these probes possess its own unique features that may yield useful information about a structure while another probe may not.
- The distance between the probe and the specimen under inspection, commonly referred to as the liftoff or standoff distance, may be manipulated as an optimization parameter for increased detection sensitivity.
- The frequency of operation may also be used as an optimization parameter.
- Signal detection may be accomplished by placing a simple and inexpensive diode detector at a strategic location inside a waveguide.
- Much of the required signal processing associated with plane-wave approach (e.g. radar techniques) are rendered unnecessary in this region.
- In the overwhelming majority of the applications there is no need for an antenna.
- Inspection systems are less bulky and cumbersome and are much more operator friendly in a sense that they may be handheld or backpack carried and battery operated.
- The required microwave power associated with these inspection systems is in the low milliwatt range.

The only aspect that may be considered negative about using this approach is that, unlike the plane-wave (far-field) approach, the formulation of the interaction of the electric and magnetic fields with complex layered composite structures in the near-field region of most probes is not simple and straight forward. In most cases a combination of analytical and numerical techniques must be used to derive a comprehensive solution for the interaction of these fields with complicated material media such as thick sandwich composites composed of many dielectric layers each with a different thickness and dielectric properties. In some cases the aperture fields (i.e. source) may be affected by the specimen under test, unlike the far-field approach. For such cases this influence must also be taken into account (i.e. inclusion of higher-order modes). However, when such solutions are obtained, then all of the features listed above are attainable, particularly the optimization feature associated with these techniques.

This chapter is primarily devoted to the use of open-ended rectangular waveguides (which is the most common near-field probes) for the inspection of multi-layered dielectric composites. Once the interaction of the fields radiated by an open-ended rectangular waveguide probe has been electromagnetically modeled, the utility of this probe for a number of practical and diverse applications will be demonstrated. Some of these results will be compared to their experimental counterparts as well.

3.4 ELECTROMAGNETIC MODELING OF THE INTERACTION OF AN OPEN-ENDED RECTANGULAR WAVEGUIDE WITH MULTI-LAYERED DIELECTRIC COMPOSITES[1]

As mentioned earlier, non-contact non-destructive evaluation of disbonds, delaminations and minute thickness variations in stratified composite materials is of great interest in many industrial applications. Good examples would be thickness variations, disbonds and delamination detection in ceramic, synthetic rubber, and honeycomb composite structures. In many instances, the layered structure is backed by a conducting plate, or it is a dielectric coating on top of a conducting sheet.

Much work has been done on rigorous theoretical analysis of a rectangular waveguide radiating into a multi-layered composite structure. The applications at which these works have been aimed at are just as diverse. Most of the earlier literature have addressed the problem of plasma covered aperture antennas (Galejs, 1965; Villeneuve, 1965; Compton, 1964; Croswell *et al.*, 1968). Rigorous analysis have also been presented mostly in application to measurement of dielectric properties of materials and field interactions with biological tissues (Teodoridis, Sphicopoulos and Gardiol, 1985; Jamieson and Rozzi, 1977; Decreton and Gardiol, 1974; MacPhie and Zaghloul, 1980; Nakita and Uzunoglu, 1989). Considering typical measurement resolutions and accuracies in near-field examination of generally lossy composite media, most rigorous numerical results offer little practical advantage when taking into account the necessary computation time and difficulties involved in replicating the results by means of well-controlled experimental setups.

The approach adopted here was originally applied by Compton (Compton, 1964) in application to radiation from plasma covered aperture antennas. It is assumed that the aperture field distribution is that of the

[1] Portions of this section are reprinted with permission from (Transactions on Microwave Theory and Techniques, MTT-42, no. 3, pp. 389-95, March 1994). © 1994 IEEE.

dominant mode. This approach was first implemented and tested for the practical case of thickness measurement and variation detection of lossy dielectric slabs backed by a conducting plate (Bakhtiari, Ganchev and Zoughi, 1993b). Here this solution is expanded to include general N-layer stratified dielectric composite media which may be terminated by a conducting sheet or by an infinite half-space of free-space or another dielectric. This is done using a Fourier transform boundary matching technique to construct the field solutions in the multi-layered medium being inspected by the open-ended rectangular waveguide. A variational expression for the terminating aperture admittance in an infinite ground plane is used to construct the desired solution.

3.4.1 Theoretical formulation

Figures 3.1 and 3.2 show two general cases of multi-layered dielectric composites being inspected by an open-ended rectangular waveguide. Variational expression for the terminating admittance of the open-ended rectangular waveguide for these two cases can be written as (Compton, 1964):

$$Y = G + jB = \frac{\iint\limits_{aperture}\left[\boldsymbol{E}(x,y,0)\times\boldsymbol{W}(x,y)\right]\cdot\boldsymbol{u}_z\,dx\,dy}{\left[\iint\limits_{aperture}\boldsymbol{E}(x,y,0)\cdot\boldsymbol{e}_o(x,y)dx\,dy\right]^2} \tag{3.1}$$

where,

$$\boldsymbol{W}(x,y) = \boldsymbol{H}(x,y,0) + \sum_{i=1}^{\infty} Y_i \boldsymbol{h}_i(x,y) \iint\limits_{aperture} \boldsymbol{E}(\eta,\xi,0)\cdot\boldsymbol{e}_i(\eta,\xi)\,d\eta\,d\xi \tag{3.2}$$

and $\boldsymbol{E}(x,y,0)$ and $\boldsymbol{H}(x,y,0)$ represent the aperture field distributions. Bold characters indicate vector quantities, and $\boldsymbol{u}_x$, $\boldsymbol{u}_y$, $\boldsymbol{u}_z$ are the coordinate system unit vectors.

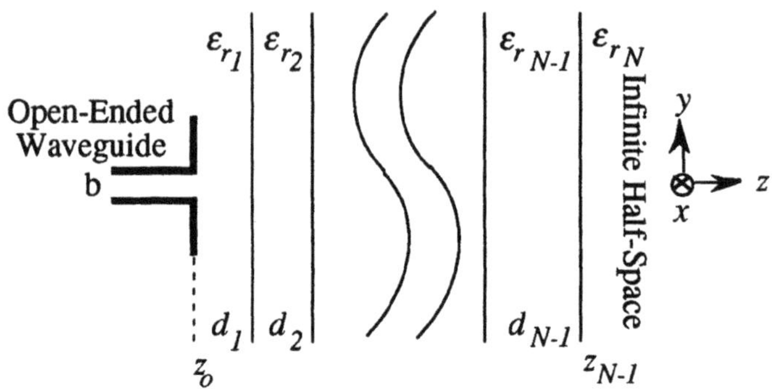

Fig. 3.1 Cross-section of an open-ended rectangular waveguide radiating into a multi-layered composite terminated by an infinite half-space (Bakhtiari *et al.*, 1994).

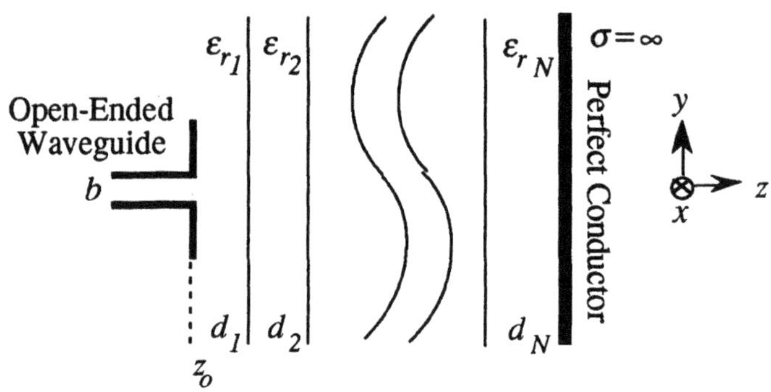

Fig. 3.2 Cross-section of an open-ended rectangular waveguide radiating into a multi-layered composite terminated by a conducting sheet (Bakhtiari *et al.*, 1994).

G and B are the aperture conductance and susceptance, respectively. This admittance expression is constructed using transverse vector mode functions and their orthogonal properties (Harrington, 1961). $\boldsymbol{e}_i$ and $\boldsymbol{h}_i$ are the ith vector mode functions and Y_i is the characteristic admittance of the ith waveguide mode. It has been shown that equation (3.1) is stationary with respect to variations of the aperture E-field about its exact value (Compton, 1964; Collin, 1960; Galejs, 1969). Thus, a reasonable approximation for the electric field results in a good estimate of the aperture admittance. It must be noted that as standard practice, a square flange with sides greater than $1\lambda_o$ is a good approximation to the theoretical assumption of an infinite flange for near-field measurements (Croswell,

Rudduck and Hatcher, 1967).

With the TE_{10} mode incident on the aperture, a normalized symmetrical aperture electric field distribution can be written as:

$$\boldsymbol{E}_y(x,y,0) = \boldsymbol{e}_o(x,y) \cdot \boldsymbol{u}_y = \begin{cases} \sqrt{\dfrac{2}{ab}} \cos\left(\dfrac{\pi x}{a}\right) & (x,y) \in Aperture \\ 0 & (x,y) \notin Aperture \end{cases} \tag{3.3}$$

where a and b are the broad and narrow dimensions of the waveguide cross-section, respectively. A Fourier transform boundary matching technique is used to construct the field solutions in an N-layer stratified generally lossy dielectric medium. The transverse field components are expanded in each layer in terms of Fourier integrals. Subsequently, appropriate boundary conditions across each interface is enforced to solve for the unknown field coefficients in each medium. As shown in Figs. 3.1 and 3.2 the waveguide is radiating into a multi-layered composite medium which may be terminated into an infinite half-space or a perfect conducting sheet, respectively. Each layer is assumed to be homogeneous and non-magnetic with relative complex dielectric property ε_{rn}.

The fields outside the waveguide may be constructed using a single vector potential with two components. In each layer, denoted by layer number n, fields must satisfy the source-free wave equation. These field components may be written as:

$$\boldsymbol{E}^n(x,y,z) = -\nabla \times \boldsymbol{\Pi}^n \tag{3.4a}$$

$$\boldsymbol{H}^n(x,y,z) = \frac{1}{j\omega\mu_o}\left[k_n^2 \boldsymbol{\Pi}^n + \nabla\nabla \cdot \boldsymbol{\Pi}^n\right] \tag{3.4b}$$

where,

$$\boldsymbol{\Pi}^n = \boldsymbol{\Phi}_n \boldsymbol{u}_x + \boldsymbol{\Psi}_n \boldsymbol{u}_y \tag{3.4c}$$

General solutions of equation (3.4a) and (3.4b) may be expressed in terms of integrations over the entire mode space as

$$E^n_{\begin{Bmatrix}x\\y\end{Bmatrix}}(x,y,z)=\frac{jk_{z_n}}{(2\pi)^2}\int_{-\infty}^{\infty}\int_{-\infty}^{\infty}\left[\mp A^+_{\begin{Bmatrix}\Psi_n\\\Phi_n\end{Bmatrix}}e^{-jk_{z_n}z}\pm A^-_{\begin{Bmatrix}\Psi_n\\\Phi_n\end{Bmatrix}}e^{jk_{z_n}z}\right]$$

$$e^{-j\left(k_x x+k_y y\right)}dk_x dk_y \qquad (3.5a)$$

$$H^n_{\begin{Bmatrix}x\\y\end{Bmatrix}}(x,y,z)=\frac{1}{(2\pi)^2}\int_{-\infty}^{\infty}\int_{-\infty}^{\infty}\frac{1}{j\omega\mu_o}\left\{\left[k_n^2-k^2_{\begin{Bmatrix}x\\y\end{Bmatrix}}\right]\right.$$

$$\left[A^+_{\begin{Bmatrix}\Phi_n\\\Psi_n\end{Bmatrix}}e^{-jk_{z_n}z}+A^-_{\begin{Bmatrix}\Phi_n\\\Psi_n\end{Bmatrix}}e^{jk_{z_n}z}\right]$$

$$\left.-k_x k_y\left[A^+_{\begin{Bmatrix}\Psi_n\\\Phi_n\end{Bmatrix}}e^{-jk_{z_n}z}+A^-_{\begin{Bmatrix}\Psi_n\\\Phi_n\end{Bmatrix}}e^{jk_{z_n}z}\right]\right\}$$

$$e^{-j\left(k_x x+k_y y\right)}dk_x dk_y \qquad (3.5b)$$

where k_n is the complex propagation constant for the nth layer and,

$$k_{z_n}=\sqrt{k_n^2-k_x^2-k_y^2} \qquad (3.6a)$$

is taken as:

$$k_{z_n}=\left|\mathrm{Re}\left\{k_{z_n}\right\}\right|-j\left|\mathrm{Im}\left\{k_{z_n}\right\}\right| \qquad (3.6b)$$

to comply with the appropriate direction of propagation. Referring to Fig. 3.1, only positively traveling waves exist in the Nth layer which is unbounded in $+z$ direction. Thus, for the field components in this region only those terms in equation (3.5a) and (3.5b) which are associated with positive direction of propagation remain. Using Fourier properties of the field components in equation (3.5) and forcing the continuity of the tangential field components at each interface, the unknown field coefficients for each layer can be determined. This is done by introducing the following normalized (with respect to k_o) parameters inside the formulation:

$$a' = k_o a \quad , b' = k_o b \quad , Z_n = k_o \sum_{i=1}^{n} d_i \quad , \mathcal{A}_I^{\pm} = k_o^2 A_I^{\pm}$$

$$\kappa_{x,y} = \frac{k_{x,y}}{k_o} \quad , \kappa_n = \frac{k_{z_n}}{k_o} = \sqrt{\varepsilon_{r_n} - \kappa_x^2 - \kappa_y^2}$$

where d_i is the thickness of the ith layer, and the index I refers to Ψ or Φ component of the vector potential. Taking inverse transform of both components of equation (3.5a) at the aperture results in the following relations between the normalized field coefficients in layer 1:

$$\mathcal{A}_{\Psi_1}^{+} = \mathcal{A}_{\Psi_1}^{-} \tag{3.7}$$

$$\mathcal{A}_{\Phi_1}^{+} - \mathcal{A}_{\Phi_1}^{-} = \frac{4\pi}{j\kappa_1}\sqrt{\frac{2a'}{b'}}\,\frac{\cos\left(\frac{\kappa_x a'}{2}\right)\sin\left(\frac{\kappa_y b'}{2}\right)}{\kappa_y\left[\pi^2 - (\kappa_x a')^2\right]} = \frac{\mathcal{G}(\kappa_x, \kappa_y)}{j\kappa_1} \tag{3.8}$$

where $\mathcal{G}$ is chosen to simplify the notation. Forcing the boundary conditions; namely, continuity of the transverse E- and H-field components across each interface results in the following set of relations between the field coefficients of adjacent layers

$$\mathcal{A}^{+}_{\begin{Bmatrix}\Psi_m\\ \Phi_m\end{Bmatrix}} e^{-j\kappa_m Z_m} - \mathcal{A}^{-}_{\begin{Bmatrix}\Psi_m\\ \Phi_m\end{Bmatrix}} e^{j\kappa_m Z_m} =$$

$$\frac{\kappa_{m+1}}{\kappa_m}\left[\mathcal{A}^{+}_{\begin{Bmatrix}\Psi_{m+1}\\ \Phi_{m+1}\end{Bmatrix}} e^{-j\kappa_{m+1} Z_m} - \mathcal{A}^{-}_{\begin{Bmatrix}\Psi_{m+1}\\ \Phi_{m+1}\end{Bmatrix}} e^{j\kappa_{m+1} Z_m}\right] \tag{3.9}$$

$$\left(\varepsilon_{r_m} - \kappa^2_{\begin{Bmatrix}x\\y\end{Bmatrix}}\right)\left[\mathcal{A}^+_{\begin{Bmatrix}\Phi_m\\\Psi_m\end{Bmatrix}} e^{-j\kappa_m Z_m} + \mathcal{A}^-_{\begin{Bmatrix}\Phi_m\\\Psi_m\end{Bmatrix}} e^{j\kappa_m Z_m}\right]$$

$$-\left(\kappa_x\kappa_y\right)\left[\mathcal{A}^+_{\begin{Bmatrix}\Psi_m\\\Phi_m\end{Bmatrix}} e^{-j\kappa_m Z_m} + \mathcal{A}^-_{\begin{Bmatrix}\Psi_n\\\Phi_m\end{Bmatrix}} e^{j\kappa_m Z_m}\right]$$

$$=\left(\varepsilon_{r_{m+1}} - \kappa^2_{\begin{Bmatrix}x\\y\end{Bmatrix}}\right)\left[\mathcal{A}^+_{\begin{Bmatrix}\Phi_{m+1}\\\Psi_{m+1}\end{Bmatrix}} e^{-j\kappa_{m+1} Z_m} + \mathcal{A}^-_{\begin{Bmatrix}\Phi_{m+1}\\\Psi_{m+1}\end{Bmatrix}} e^{j\kappa_{m+1} Z_m}\right]$$

$$-\left(\kappa_x\kappa_y\right)\left[\mathcal{A}^+_{\begin{Bmatrix}\Psi_{m+1}\\\Phi_{m+1}\end{Bmatrix}} e^{-j\kappa_{m+1} Z_n} + \mathcal{A}^-_{\begin{Bmatrix}\Psi_{m+1}\\\Phi_{m+1}\end{Bmatrix}} e^{j\kappa_{m+1} Z_m}\right] \tag{3.10}$$

where $m = 1,2,\ldots,(N\text{-}1)$ is the layer number. Consequently, when the medium is terminated into an infinite half-space the contribution of negative traveling wave components is omitted in equations (3.9) and (3.10). For a layered media terminated by a perfectly conducting sheet, the vanishing tangential E-field components over the conductor surface renders an additional set of relations between the field coefficients in the Nth layer as:

$$\mathcal{A}^-_{\begin{Bmatrix}\Psi_N\\\Phi_N\end{Bmatrix}} = e^{-j2\kappa_N Z_N}\,\mathcal{A}^+_{\begin{Bmatrix}\Psi_N\\\Phi_N\end{Bmatrix}} \tag{3.11}$$

Solving simultaneously a system of $4N$-2 equations for the N-layer case terminated by an infinite half-space results in the sought for field coefficients. Similarly, a system of $4N$ equations has to be solved for the case of N-layer medium terminated by a conducting sheet. Application of Parseval's theorem allows the construction of the numerator of admittance expression in equation (3.1) as:

$$\int_{-\infty}^{\infty}\int_{-\infty}^{\infty}\left[\boldsymbol{E}(x,y,0)\times\boldsymbol{W}(x,y,0)\right]\cdot\boldsymbol{u}_z\,dxdy=$$

$$\frac{1}{(2\pi)^2}\int_{-\infty}^{\infty}\int_{-\infty}^{\infty}\mathcal{E}_y^{1*}(k_x,k_y)\mathcal{H}_x^1(k_x,k_y)\,dk_xdk_y \qquad (3.12)$$

where the $\mathcal{E}$ and $\mathcal{H}$ in the right hand side of equation (3.12) are the transformed field components of layer 1 from equations (3.5a) and (3.5b) and (*) denotes complex conjugate. Upon calculation of the field coefficients in layer 1 and the use of equation (3.12) the following admittance expression results from equation (3.1):

$$Y_n=\frac{j}{(2\pi)^2}\int_{-\infty}^{\infty}\int_{-\infty}^{\infty}\left\{\left[\varepsilon_{r_1}-\kappa_x^2\right]\left[2\mathcal{A}_{\Phi_1}^{+}+\frac{j\mathcal{G}(\kappa_x,\kappa_y)}{\kappa_1}\right]-2\kappa_x\kappa_y\mathcal{A}_{\Psi_1}^{+}\right\}$$

$$\mathcal{G}(\kappa_x,\kappa_y)\,d\kappa_x d\kappa_y \qquad (3.13)$$

where subscript n denotes normalization with respect to the free-space characteristic admittance. By applying a polar coordinate transformation of the form:

$$\kappa_x=\mathcal{R}\cos\theta\,,\ \kappa_y=\mathcal{R}\sin\theta \qquad (3.14)$$

with transformation Jacobian $\mathcal{R}$ to equation (3.13), which has two infinite integrals, the terminating aperture admittance may be evaluated by a double integral having only one unbounded range of integration as ultimately given by equation (3.15):

$$y_s=g_s+jb_s=\frac{j2}{(2\pi)^2\sqrt{1-\left(\dfrac{\pi}{a'}\right)^2}}$$

$$\times\int_{\mathcal{R}=0}^{\infty}\int_{\theta=0}^{\pi}\chi\left(\mathcal{G}(\mathcal{R},\theta)\,\mathcal{R}\,d\theta\,d\mathcal{R}\right) \qquad (3.15)$$

where,

$$\chi = \left[\varepsilon_{r_1} - (\mathcal{R}\cos\theta)^2\right]\left(2\mathcal{A}^+_{\Phi_1} + \frac{jG(\mathcal{R},\theta)}{\sqrt{\varepsilon_{r_1} - \mathcal{R}^2}}\right)$$

$$-2\mathcal{R}^2 \sin\theta\cos\theta\mathcal{A}^+_{\Psi_1} \qquad (3.16)$$

and,

$$G(\mathcal{R},\theta) = \sqrt{\frac{2a'}{b'}}\,\frac{(4\pi)\cos\left(\dfrac{a'\mathcal{R}\cos\theta}{2}\right)\sin\left(\dfrac{b'\mathcal{R}\sin\theta}{2}\right)}{(\mathcal{R}\sin\theta)\left[\pi^2 - (a'\mathcal{R}\cos\theta)^2\right]} \qquad (3.17)$$

with $\mathcal{R}$ and θ as the new variables of integration given by equation (3.14) and a' and b' are the normalized waveguide dimensions. Lossless stratified media is not the focus of this work, however, for such a case one encounters singularities in equation (3.14). This can be resolved by contour integration around singular points of the integrand located on the real axis and take into account the residue. In application to a composite medium which contains generally lossy layers, which is the case addressed here, the poles of the integrand move off the real axis and the integrand becomes smooth. This allows quick and efficient numerical integration schemes such as a Gauss quadrature method (used here) to be applied.

3.4.2 Verification of derivations[2]

The complex reflection coefficient, Γ, is related to the normalized complex admittance, y_s, by $\Gamma = (1 - y_s)/(1 + y_s)$. Thus, the magnitude and phase of Γ; namely, $|\Gamma|$ and ϕ can be calculated and subsequently measured as a function of various parameters associated with the multi-layered composites shown in Figs. 3.1 and 3.2. There are several techniques for measuring the magnitude and phase of Γ. Using a short-circuited waveguide method (for calibration), they can be directly measured from *SWR* and shift in the location of the standing wave null compared to a short-circuited waveguide (Ganchev, Bakhtiari and Zoughi, 1992). Another approach involves utilizing a calibrated vector network analyzer such as the HP8510 or the HP8720 series. A third approach involves the use of a dual-arm microwave reflectometer or bridge (Bakhtiari, Ganchev and Zoughi, 1993a). A fourth

[2] Portions of this sub-section are reprinted with permission from (Transactions on Instrumentation and Measurement, IM-42, no. 1, pp. 19-24, February 1993). ©1993 IEEE.

approach involves designing and building single reflectometers, phase detectors or dual reflectometers which are relatively easy and inexpensive to design and construct. Such systems may then be used for comparative measurements or can be calibrated to render magnitude and phase of the reflection coefficient referenced to the waveguide aperture. For most practical non-destructive applications it is more important to obtain information about changes in the structure of a composite as opposed to information about its absolute geometrical status.

The results of this derivation were compared to those obtained by the Lewin's infinite half-space formulation (Lewin, 1951). Table 3.1 shows the results at a frequency of 10 GHz for several dielectric materials. Clearly, there is good agreement between the two approaches. For more results of equation (3.15) the reader is referred to (Bakhtiari, 1992).

Table 3.1 Comparison of the results of the calculated admittance at the waveguide aperture using the present and Lewin's derivations at a frequency of 10 GHz (Bakhtiari, Ganchev and Zoughi, 1993b).

ε'_r	$\tan\delta$	equation (3.15)	Lewin's
2.6	0.01	2.074 + j0.555	2.068 + j0.556
7.0	0.10	3.441 + j0.062	3.432 + j0.062
10.5	0.238	4.166 - j0.166	4.166 - j0.166

3.4.3 Thickness determination of dielectric sheets backed by conducting plates[3]

One of the areas of non-destructive testing and evaluation that this technique can be effectively applied to is absolute thickness determination and thickness variation detection of dielectric coatings such as paint, primer, thermal barrier coatings, synthetic rubber laminate coatings all backed by a conducting plate. Figure 3.3a shows the general schematic of these applications, while Fig. 3.3b shows the incorporation of a standoff distance in this geometry.

To investigate the influence of dielectric sheet thickness on the normalized admittance of the waveguide, variation of g_s and b_s as a function of the thickness of two types dielectric materials; namely, Plexiglass ($\varepsilon'_r = 2.59$, $\tan\delta = 0.007$) and carbon loaded rubber ($\varepsilon'_r = 7.25$, $\tan\delta = 0.103$) are shown in Figs. 3.4 and 3.5. The change in the sign of b_s indicates that at some thicknesses the aperture is terminated by a capacitive

[3] Portions of this sub-section are reprinted with permission from (Transactions on Instrumentation and Measurement, IM-42, no. 1, pp. 19-24, February 1993) © 1993 IEEE, and with permission from *Materials Evaluation*, vol. 51, no. 6, pp. 740-43, Copyright © 1993 The American Society for Nondestructive Testing, Inc.

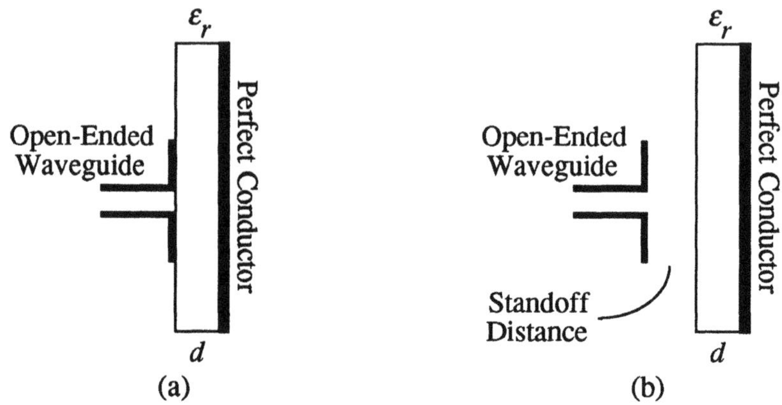

Fig. 3.3 The general schematic of a) a dielectric coating backed by a conducting plate, b) incorporation of a standoff distance in the geometry.

or inductive load.

There is no conductance or susceptance measuring equipment at microwave frequencies. Thus, in order to be able to apply this technique in practice, measurable parameters, as described earlier must be utilized. The reflection coefficient at the waveguide aperture can be related to its admittance as mentioned in the previous section. Several measurable parameters, depending on the measurement method, may then be related to the reflection coefficient. When using a slotted waveguide section terminated by the conducting plate-backed dielectric coating, the shift in the location of standing wave minimum and the corresponding *SWR*, both of which are directly related to Γ, may be used as measurable parameters (see Chapter 2). The shift in the location of the minimum is related to the phase of the reflection coefficient and the *SWR* is related to the magnitude of reflection coefficient. As explained in Chapter 2, using a calibrated attenuator eliminates the measurement dependency on the properties of the detector diode. Therefore, one may also use the attenuation factor as a direct measurable parameter (Attenuation (dB) = 20log(*SWR*)).

It is important to note that in practice usually a signal related to the phase or magnitude of reflection coefficient may also be used as an indicator of the absolute thickness or the thickness variation. When dealing with thin and low-loss dielectric materials, information about the phase of reflection coefficient is much more useful than that about the magnitude of reflection coefficient. However, when dealing with lossy dielectric materials (thin or thick) the magnitude information may also be used as an effective parameter for thickness variation determination, as will be seen later.

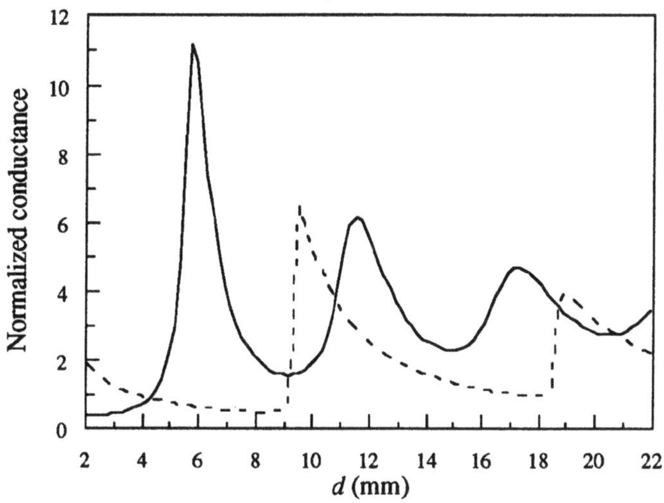

Fig. 3.4 Calculated variations of the normalized conductance, g_s, as a function of thickness for Plexiglass (dashed line) and carbon loaded rubber (solid line) at a frequency of 10 GHz (Bakhtiari, Ganchev and Zoughi, 1993b).

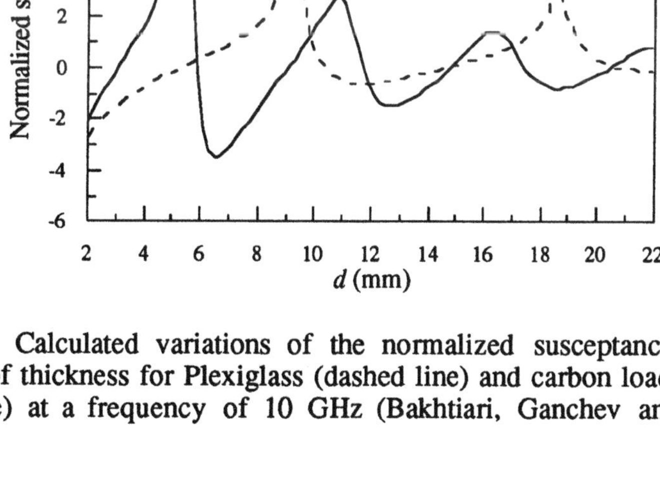

Fig. 3.5 Calculated variations of the normalized susceptance, b_s, as a function of thickness for Plexiglass (dashed line) and carbon loaded rubber (solid line) at a frequency of 10 GHz (Bakhtiari, Ganchev and Zoughi, 1993b).

Figures 3.6 and 3.7 show the comparison between the calculated and measured attenuation and phase for a Plexiglass sheet, with $\varepsilon'_r = 2.59$, $\tan\delta = 0.007$ backed by a conducting plate, as a function of the sheet thickness at a frequency of 10 GHz. Figures 3.8 and 3.9 show similar results for the carbon loaded rubber with ($\varepsilon'_r = 12.6$, $\tan\delta = 0.19$) at the same frequency.

The results reveal several important facts. For small thicknesses of both of these dielectric materials, one may use both the phase and attenuation to detect thickness variation. When using phase, small thickness variations cause significant phase change. Therefore, this technique is clearly quite sensitive to minute thickness variation, and hence may be used a sensitive and accurate thickness measuring gauge. The results also indicate that the measurement sensitivity decreases as the thickness of the coating increases for a given frequency of operation. In other words, when the electrical thickness of the dielectric slab increases towards an infinite half-space situation, the measurement sensitivity decreases.

Although the above results primarily demonstrate the ability of this approach to detect dielectric coating thickness variation, one may also use this technique for absolute thickness determination. To calculate the thickness of a coating (given its dielectric properties), the conductance and susceptance may be measured (via the measurement of Γ) and using a root finding scheme the thickness may be determined (Ganchev, Bakhtiari and Zoughi, 1992). Starting with initial lower and upper bounds for thickness, g_s and b_s may be iterated alternately to find the root which closely estimates the thickness of the dielectric coating. For both, the root is found within some prescribed range which may be initially estimated from plots of the variations of these parameters over some large interval. This also helps in reducing calculation time since a finer range can be estimated in advance for the position of the root. As the loss tangent of the dielectric sheet decreases, the integration in equation (3.15) requires finer increments to be evaluated due to slower convergence. This simply means that the integrand approaches the singular behavior which eventually appears for lossless dielectric coatings. As the loss tangent increases the integrand becomes more smooth and the integration becomes faster. Hence, one may use a Gauss quadrature method with a fixed number of intervals.

Using the above approach, thicknesses of two carbon black loaded rubber sheets with $\varepsilon'_r = 7.25$ and $\tan\delta = 0.103$ and with $\varepsilon'_r = 12.6$ and $\tan\delta = 0.19$ were measured at a frequency of 10 GHz (Bakhtiari, Ganchev and Zoughi, 1993b). Tables 3.2 and 3.3 show the results of measurements and calculations of the admittance at the waveguide aperture for both of these materials, respectively. Measured thickness refers to the actual thickness of the rubber sheets measured using a micrometer. The results show that for both rubber samples the measurement error for the range of thicknesses measured is less than 3%.

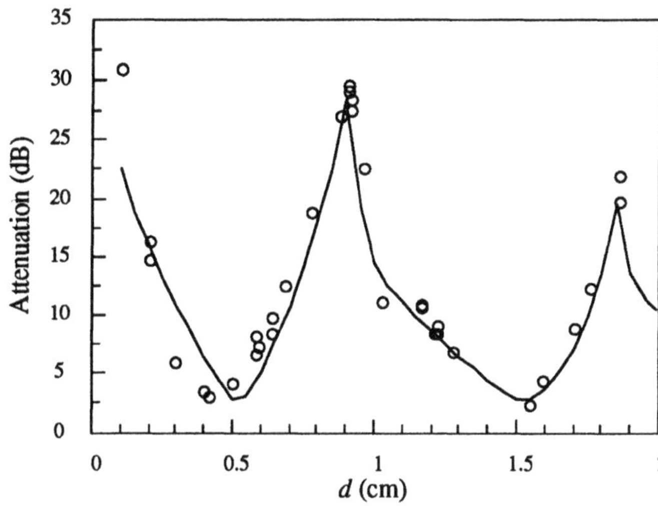

Fig. 3.6 Comparison of the calculated (solid line) and measured (circles) attenuation as a function of thickness at a frequency of 10 GHz for Plexiglass with $\varepsilon'_r = 2.59$, $\tan\delta = 0.007$ (Bakhtiari, Ganchev and Zoughi, 1993b).

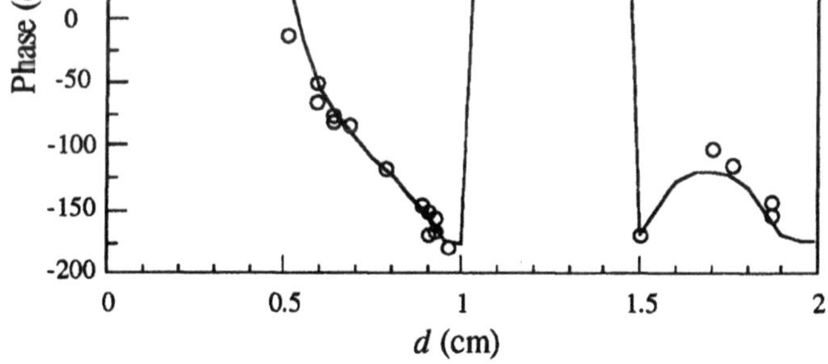

Fig. 3.7 Comparison of the calculated (solid line) and measured (circles) phase as a function of thickness at a frequency of 10 GHz for Plexiglass with $\varepsilon'_r = 2.59$, $\tan\delta = 0.007$ (Bakhtiari, Ganchev and Zoughi, 1993b).

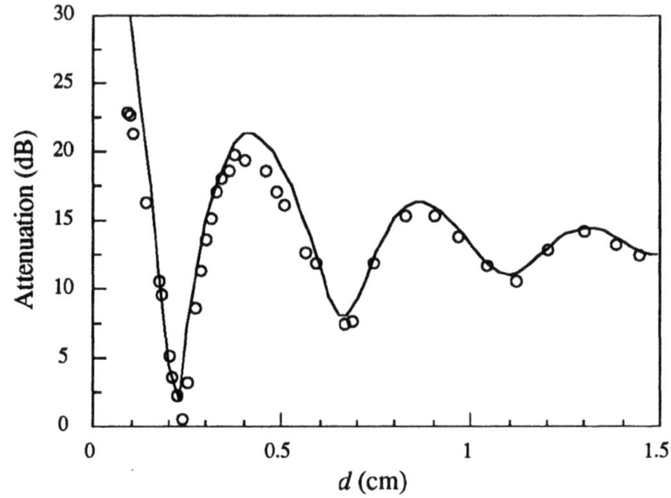

Fig. 3.8 Comparison of the calculated (solid line) and measured (circles) attenuation as a function of thickness at a frequency of 10 GHz for carbon loaded rubber with $\varepsilon'_r = 12.6$, $\tan\delta = 0.19$ (Bakhtiari, Ganchev and Zoughi, 1993b).

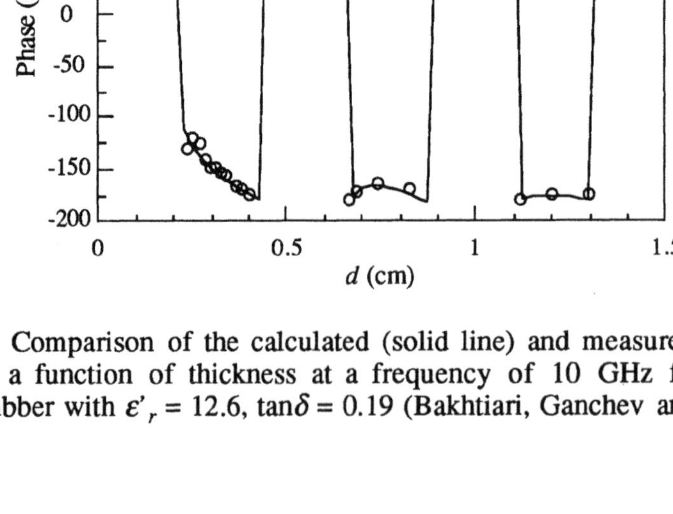

Fig. 3.9 Comparison of the calculated (solid line) and measured (circles) phase as a function of thickness at a frequency of 10 GHz for carbon loaded rubber with $\varepsilon'_r = 12.6$, $\tan\delta = 0.19$ (Bakhtiari, Ganchev and Zoughi, 1993b).

Table 3.2 Thickness measurement results for carbon back loaded rubber with ε'_r = 7.25 and $\tan\delta$ = 0.103 at a frequency of 10 GHz (Bakhtiari, Ganchev and Zoughi, 1993b).

Measured thickness (mm)	*Measured admittance, Y*	*Calculated thickness (mm)*	*% Mean error*
3.91 ± 0.04	0.70 + j1.43	3.90 ± 0.05	0.26
5.92 ± 0.04	10.3 + j1.61	5.83 ± 0.05	1.52
7.77 ± 0.05	2.37 - j1.21	7.80 ± 0.10	0.40
11.5 ± 0.05	5.57 + 1.85	11.38 ± 0.19	1.56

Table 3.3 Thickness measurement results for carbon back loaded rubber with ε'_r = 12.6 and $\tan\delta$ = 0.19 at a frequency of 10 GHz (Bakhtiari, Ganchev and Zoughi, 1993b).

Measured thickness (mm)	*Measured admittance, Y*	*Calculated thickness (mm)*	*% Mean error*
3.63 ± 007	3.67 + j4.25	3.59 ± 0.02	0.1
3.76 ± 0.05	4.85 + j4.83	3.76 ± 0.02	0.1
5.90 ± 0.07	2.80 + j1.69.	5.89 ± 0.06	0.2
7.43 ± 0.06	3.08 + j1.64	7.45 ± 0.05	0.3
8.20 ± 0.05	3.88 + j2.55	8.03 ± 0.05	2.2

It is also important to consider the influence of frequency for a given coating dielectric properties and thickness. This issue is referred to as "measurement optimization". The proper choice of the operating frequency can significantly enhance the detection sensitivity to slight thickness variations of a given coating. To show the important role that the operating frequency plays several cases are considered here. For all of these cases, the calculated results are compared to their measured counterparts. The measured results are obtained using a dual-arm microwave reflectometer (or calibrated bridge) as shown in Fig. 3.10 (Bakhtiari, Ganchev and Zoughi, 1993a). The measurement apparatus is basically a dual-arm microwave reflectometer. The sweep oscillator generates the swept frequency signal which is then fed into a waveguide Tee. The Tee splits the input signal into two equal (phase and magnitude) signals each entering one of the reflectometer arms. In the upper arm the signal is fed into an open-ended waveguide which is in contact with the conductor backed dielectric coating. The reflected signal subsequently is coupled into the attached directional coupler and finally into the test port of a microwave network analyzer (HP8410). The other half of the input signal enters the lower arm and is reflected off of a sliding short circuit. The reflected signal in this arm is then fed into the reference port of the network analyzer for comparison (phase and magnitude) with the test signal. The sliding short in the reference arm allows compensation for the extra length of the waveguide in the test arm over the entire range of frequency sweep. Once the device is calibrated with respect to a short circuit at the test port, any detected change in the measurement parameters is as a result of changes in the dielectric medium. The use of a computer controller, via an A/D board, allows real time processing and display of the measurement results over the swept bandwidth.

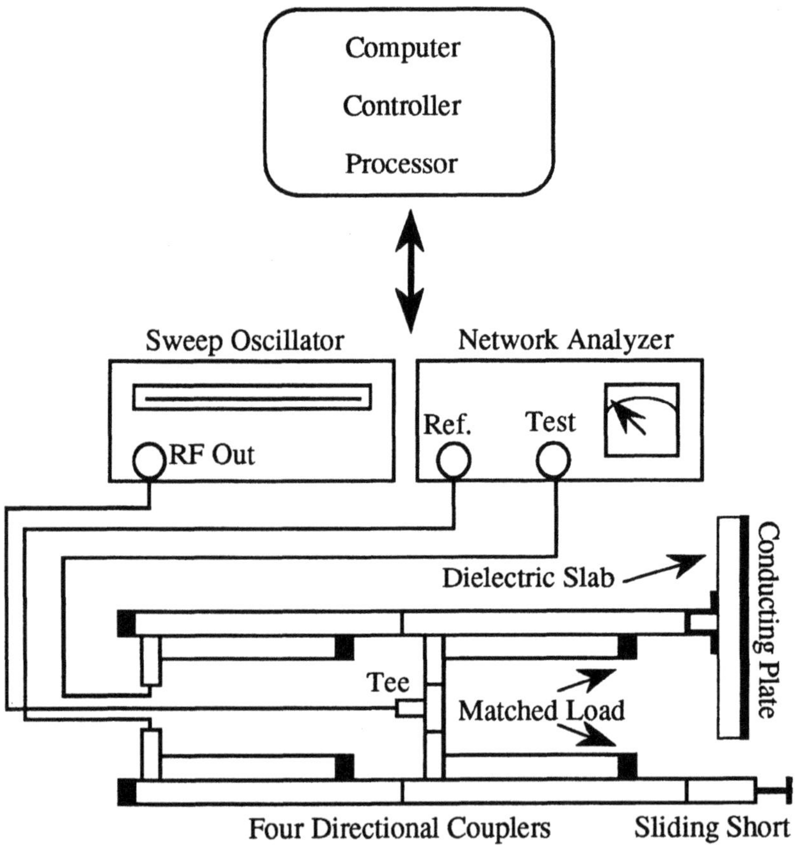

Fig. 3.10 Dual-arm reflectometer apparatus (Bakhtiari, Ganchev and Zoughi, 1993a).

To better understand the measurement resolution and sensitivity due to small variations in a dielectric sheet thickness, a 2 mm-thick coating of a material commonly used as thermal barrier coating in extremely high temperature environments is used. This coating on top of a conducting plate is considered in the category of low-loss and high-permittivity ceramics having a measured $\varepsilon_r = 20 - j0.02$ at 9 GHz. Figure 3.11 shows the results of the calculated and measured phase of reflection coefficient from 8.2 to 9.5 GHz. The solid line is the theoretical predication for a coating thickness of 2 mm. The circles show the result of the swept frequency measurement. Clearly, the two results are in fairly good agreement. The deviation of the measured results from the calculated results is attributed to the nonlinearities in the measurement apparatus, slight variation in the dielectric properties of the material in the swept frequency range, and partially to the calibration process. Very good calibration can be achieved for a single frequency, however, over a wide band of frequencies the calibration quality becomes poorer. This problem may be remedied by post processing of the measurement data or the use of digital microwave network analyzers such as the HP8510. The theoretical results for a ±50 micrometer change (2.5% variation) in the thickness of this coating are also obtained, and are shown as the dashed lines in Fig. 3.11. The results indicate that at frequencies higher than 9.5 GHz the resolution for detecting the thickness variation is relatively poor. However, in the range of 8.2-8.6 GHz, a 50 micrometer thickness change causes about 25° of phase change. This translates to a resolution of about 2 micrometers per degree (although this is not always a linear relationship). It is important to note that the phase can be measured accurately to within one degree. Clearly, the choice of operating frequency strongly affects the measurement resolution. From Fig. 3.11 it is also evident that frequencies lower than 8.2 GHz may render higher measurement sensitivity (the measurement apparatus here was limited to X-band region, 8.2-12.4 GHz). The most important conclusion that may be drawn from this analysis is that resolutions (or measurement sensitivity to thickness variation) in the few micrometer range can be achieved even at relatively low microwave frequencies. For plane-wave approach, this type of resolutions are thought to be associated with operating frequencies in the upper regions of the millimeter wave range.

Figure 3.12 shows the theoretical and experimental results for two samples of carbon loaded rubber with $\varepsilon_r = 12.4 - j2.4$ with thicknesses of 2.08 mm and 2.18 mm respectively, placed on top of a conducting plate. Once again, the calculated results are used to deduce information about the appropriate range of frequency in the X-band region which would result in maximum thickness measurement sensitivity. The results show significant difference between the two phases, both calculated and measured, due to 0.1 mm (100 micrometers) change in the material thickness. Calculation of the phase difference between the two samples at around 10.3 GHz (about 150°) indicates variations of less than 1 micrometers per degree, as shown in Fig. 3.13.

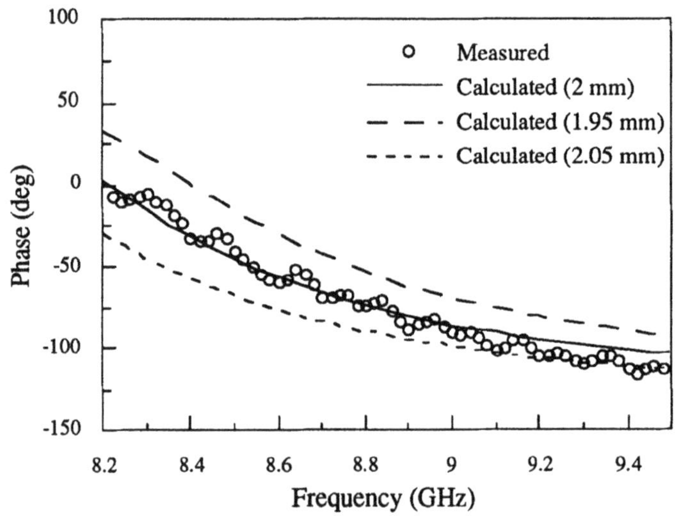

Fig. 3.11 Swept frequency comparison of the calculated and measured results for the phase of reflection coefficient for a low-loss dielectric ceramic (thermal barrier coating) (Bakhtiari, Ganchev and Zoughi, 1993a).

The effect of permittivity variations on the phase of reflection coefficient may also be investigated in a similar manner. In other words, if the dielectric properties of a material is sought, instead of measuring the dielectric properties directly (as described in Chapter 2), one may measure the phase and magnitude of reflection coefficient for a certain thickness of that material whose thickness is precisely measured. Therefore, this method may also be considered when dielectric material characterization is of interest.

This microwave non-destructive technique for accurate thickness measurement or dielectric property monitoring of dielectric materials backed by a conducting plate shows tremendous promise. The interaction of a microwave signal radiated by an open-ended rectangular waveguide into a conductor backed dielectric sheet is clearly a complicated phenomenon. To obtain maximum thickness measurement sensitivity or resolution, the choice of the operating frequency depends on the dielectric properties of the sheet and its thickness in a manner that is difficult to simply guess for. The results of the thickness measurement using the approach outlined here show that at some frequencies measurement accuracy of better than one micrometer is possible. Whereas, at a slightly different frequency in the same microwave band the measurement accuracy is significantly degraded.

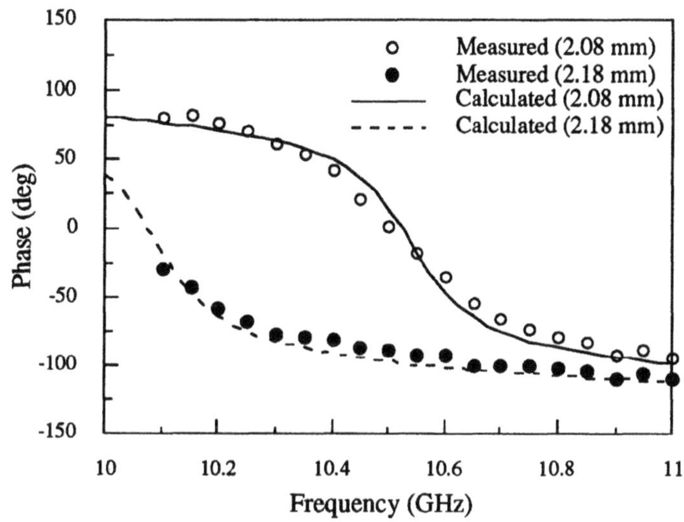

Fig. 3.12 Comparison of the calculated and measured phase of reflection coefficient for a carbon black loaded dielectric sheet with ε_r = 12.4 - j2.4 and two thicknesses of 100 micrometers apart (Bakhtiari, Ganchev and Zoughi, 1993a).

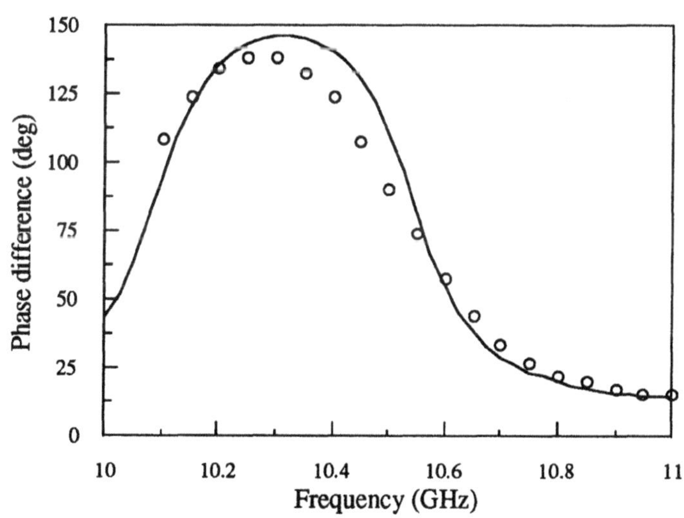

Fig. 3.13 Phase difference between the calculated (solid line) and measured (circles) results shown in Fig. 3.12 (Bakhtiari, Ganchev and Zoughi, 1993a).

The results also indicate that the detection of thickness variation in the order of a few micrometers is quite possible. This is important as these measurements were conducted in the 8.2-12.4 GHz region, and not in the millimeter wave region.

3.4.4 Stratified dielectric composite inspection[4]

There are many situations in practice that involve the inspection of multi-layered composites backed by an infinite half-space or a conducting plate as shown in Figs. 3.1 and 3.2. In addition, a dielectric material with a delamination or void, a stratified medium with a disbond, and rust under paint can be modeled as a multi-layered medium. Finally, when there is a liftoff or standoff distance incorporated in a measurement the medium falls under this category. In this section several such practical cases will be presented.

To illustrate the influence of standoff distance, consider the geometry depicted in Fig. 3.3b. Figure 3.14 shows the phase of reflection coefficient for a dielectric material with $\varepsilon_r = 7 - j0.5$ with two thicknesses of 1 mm and 1.1 mm at a frequency of 24 GHz as a function of standoff distance. Clearly, the two phase characteristics are different for the 100 micrometer thickness difference. This indicates the ability of distinguishing between these two thicknesses. In addition, Fig. 3.15 shows the phase difference between these two cases. Besides the fact that any phase difference other than zero is indicative of the thickness difference between these two dielectric sheets, the level of this difference indicates the sensitivity (resolution) by which this thickness difference is distinguished. For some standoff distance ranges (i.e. 1.5-3 mm) the difference is high enough and relatively constant. The practical ramification in this region is that some change in the standoff distance does not hinder the detection significantly, and still yields high degree of measurement sensitivity. On the other hand, at a standoff distance of 5.8 mm, the detection sensitivity is maximum (60° for 100 micrometers of thickness change). However, slight variations in the standoff distance of around 5.8 mm causes significant change in the phase difference. Therefore, when such high detection sensitivities are required, one may use a mechanical set up to hold the standoff distance constant about this value, or use an electronic mechanism to compensate for small variations in standoff distance. On the contrary, when operating at standoff distances of 4.8 mm, 6.9 mm, 10.7 mm and 13.9 mm the calculated phase difference is zero and hence no detection of thickness difference is

[4] Portions of this sub-section are reprinted with permission from (Transactions on Microwave Theory and Techniques, MTT-42, no. 3, pp. 389-95, March 1994). © 1994 IEEE.

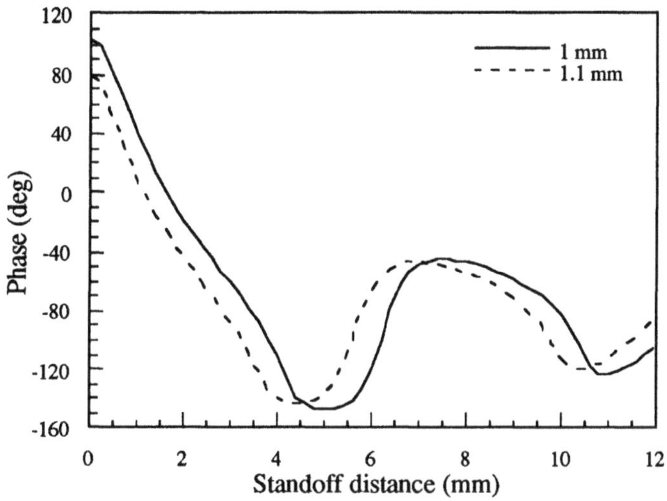

Fig. 3.14 Comparison of the calculated phase of reflection coefficient for a dielectric sheet with ε_r = 7 - j0.5 and two different thicknesses at a frequency of 24 GHz.

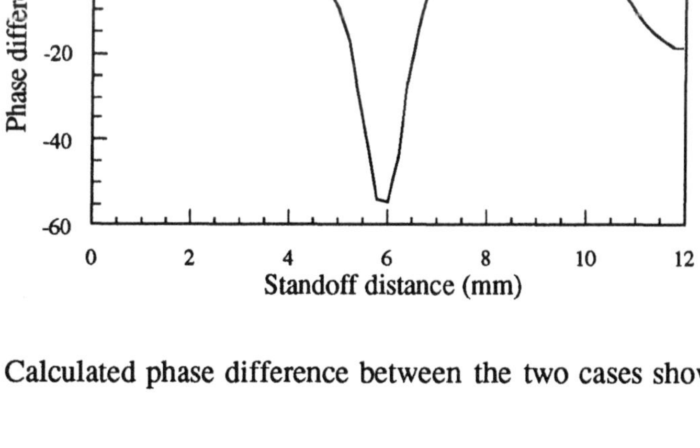

Fig. 3.15 Calculated phase difference between the two cases shown in Fig. 3.14.

possible. Finally, although not shown, one may choose to operate at two different standoff distances, such that the sign of the phase difference about the desired thickness would indicate thickness increase or decrease in the dielectric sheet. In the following sections, more discussions regarding the influence of standoff distance, which is a ramification of near-field measurements, will be presented.

Next the influence of standoff distance on a single slab terminated into free-space is studied. The geometry of this configuration is shown in Fig. 3.16. This configuration illustrates a typical two-layer model backed by free-space.

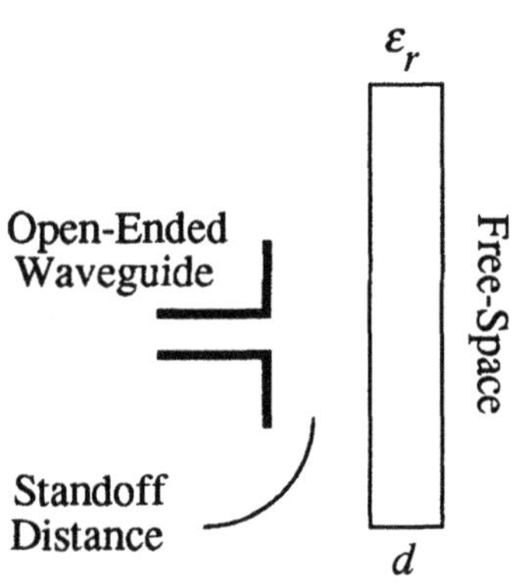

Fig. 3.16 Schematic of an open-ended rectangular waveguide, incorporating a standoff distance, radiating into a dielectric slab backed by free-space.

A uniformly thick and homogeneous dielectric slab from the family of carbon loaded rubber was used with a measured $\varepsilon_r = 8.4 - j0.9$ at 10 GHz (Bakhtiari, Qaddoumi, Ganchev, *et al.*, 1994). The mean value of the slab thickness was measured to be 7.55 mm. The results of the theoretical and the measured (using a slotted waveguide approach) attenuation and phase as a function of standoff distance, at a frequency of 10 GHz, are shown in Figs. 3.17 and 3.18. Clearly, there is good agreement between the measured and the calculated results. More importantly, the results show that the variation of attenuation and phase of reflection coefficient as a function of standoff distance (i.e. near-field measurements) are not only significant but also impossible to predict without using a rigorous electromagnetic model such as the one outlined in this chapter.

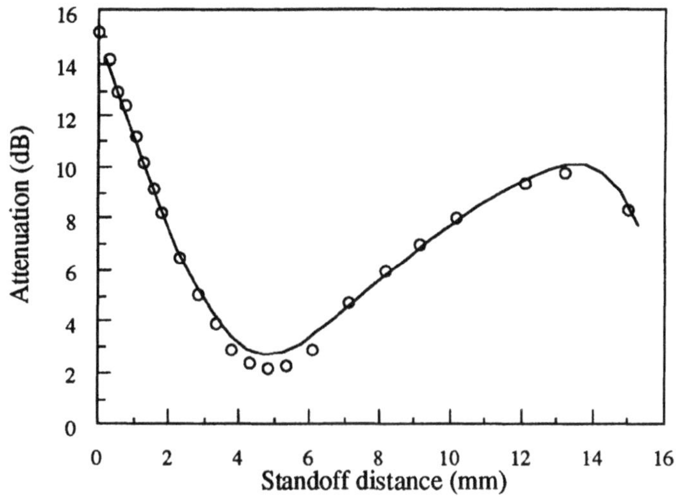

Fig. 3.17 Comparison of the calculated (solid line) and measured (circles) attenuation for a dielectric sheet with $\varepsilon_r = 8.4 - j0.9$ at 10 GHz as a function of standoff distance (Bakhtiari *et al.*, 1994).

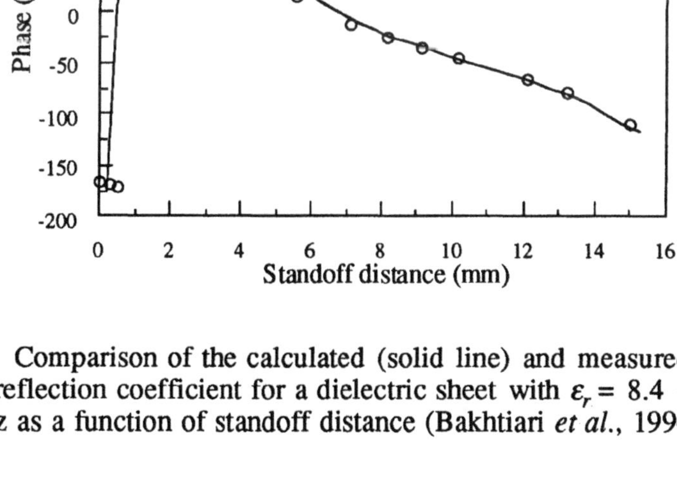

Fig. 3.18 Comparison of the calculated (solid line) and measured (circles) phase of reflection coefficient for a dielectric sheet with $\varepsilon_r = 8.4 - j0.9$ and at 10 GHz as a function of standoff distance (Bakhtiari *et al.*, 1994).

Next, variations of attenuation and phase of reflection coefficient were investigated as a function of the sheet thickness for the same rubber material with $\varepsilon_r = 8.4 - j0.9$, for a fixed arbitrary standoff distance of 5 mm and at frequencies of 10 and 24 GHz for the geometry shown in Fig. 3.16. Figures 3.19 and 3.20 show the numerical results at these frequencies as well as the measured results at 10 GHz. The measured phase at 10 GHz shows significant variation over the dielectric slab thickness range of 3-8 mm which is also predicted by the numerical results (Bakhtiari, Qaddoumi, Ganchev, *et al.*, 1994).

Detection of disbond in stratified dielectric composites is also an important practical issue. The most simple and general case of a disbonded layered composite is shown in Fig. 3.21. This constitutes a four-layer model backed by free-space. The dielectric properties of the disbond is assumed to be equal to those of free-space.

To examine the influence of disbond thickness variation, in the composite medium shown in Fig. 3.21, an experiment was conducted in which the attenuation and phase of reflection coefficient were calculated and measured, for a fixed standoff distance of 5 mm, as a function of disbond thickness between two similar dielectric sheets with $\varepsilon_r = 8.4 - j0.9$ and thicknesses of $d_1 = 5.15$ mm and $d_2 = 7.55$ mm, respectively at a frequency of 10 GHz (Bakhtiari, Qaddoumi, Ganchev, *et al.*, 1994). Figures 3.22 and 3.23 show the calculated and the measured results. The measured results, which are in good agreement with the calculations, clearly indicate that for this specific case, the phase of reflection coefficient shows much more sensitivity to the disbond thickness variation than does the attenuation. The phase drops around 15° degrees over the first 0.5 mm change in the disbond thickness. Assuming that robust phase measurement to within one degree may be obtained, then very thin disbonds (in the few micrometer range) in this composite may be detected.

One of the most common and practically critical situations in which a disbond may occur is between a dielectric sheet and a conducting plate, as shown in Fig. 3.24. This case constitutes a three-layer medium backed by a conducting plate.

Similar calculations and measurements were conducted for this case using the same dielectric sheet with $\varepsilon_r = 8.4 - j0.9$ and a thickness of d = 7.55 mm at a standoff distance of 5 mm and a frequency of 10 GHz. The calculated and the measured results for attenuation and phase of reflection coefficient are shown in Figs. 3.25 and 3.26. Both the measured attenuation and phase values, which are in close agreement with the calculated results, display significant change over the range of disbond thickness for this case. The results also indicate that the presence of the conducting sheet enhances the disbond detection sensitivity.

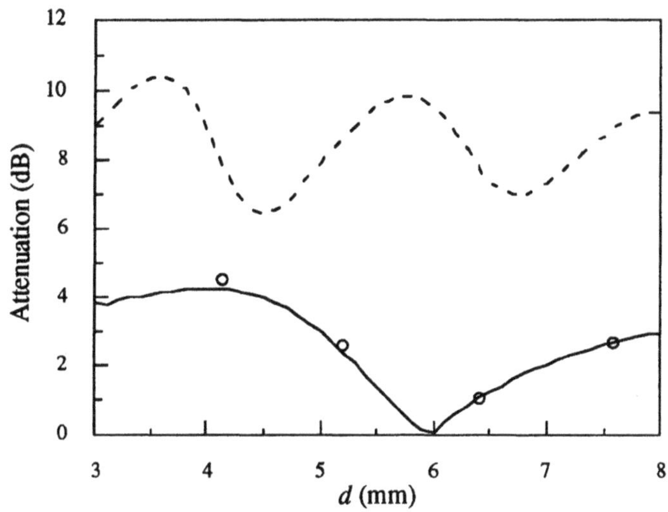

Fig. 3.19 Comparison of calculated attenuation at 10 GHz (solid line) and 24 GHz (dashed line) along with the measured attenuation at 10 GHz (circles) as a function of dielectric sheet thickness (Bakhtiari *et al.*, 1994).

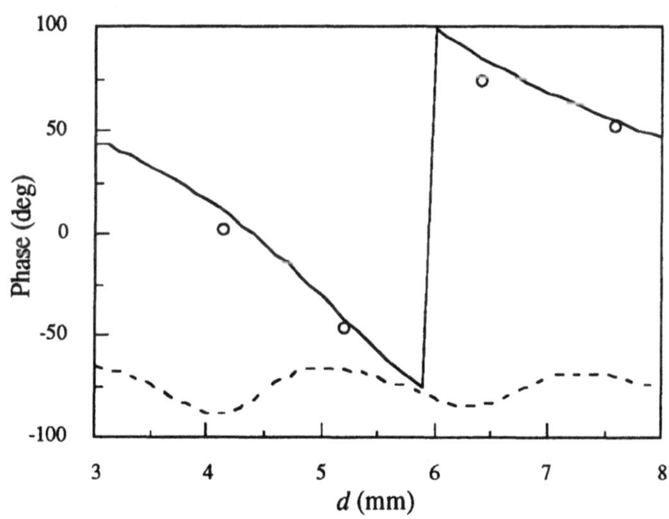

Fig. 3.20 Comparison of the calculated phase of reflection coefficient at 10 GHz (solid line) and 24 GHz (dashed line) along with the measured phase at 10 GHz (circles) as a function of dielectric sheet thickness (Bakhtiari *et al.*, 1994).

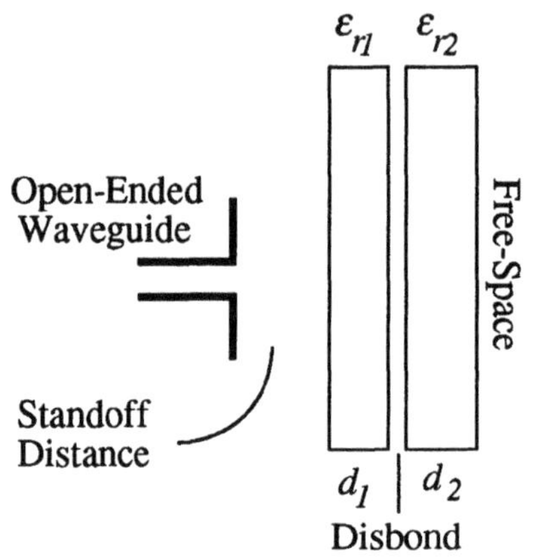

Fig. 3.21 Schematic of an open-ended rectangular waveguide, incorporating a standoff distance, radiating into a two-layer dielectric composite with a disbond in between the two layers, backed by free-space.

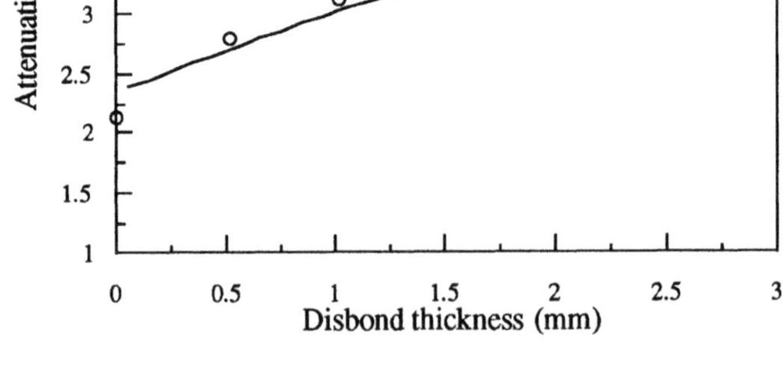

Fig. 3.22 Comparison of the calculated (solid line) and measured attenuation (circles) at 10 GHz, as a function of disbond thickness for the composite shown in Fig. 3.21 (Bakhtiari *et al.*, 1994).

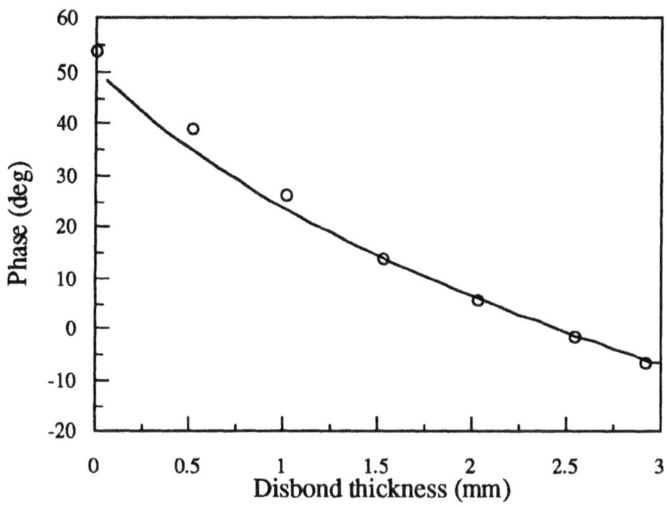

Fig. 3.23 Comparison of the calculated (solid line) and measured (circles) phase of reflection coefficient at 10 GHz, as a function of disbond thickness for the composite shown in Fig. 3.21 (Bakhtiari *et al.*, 1994).

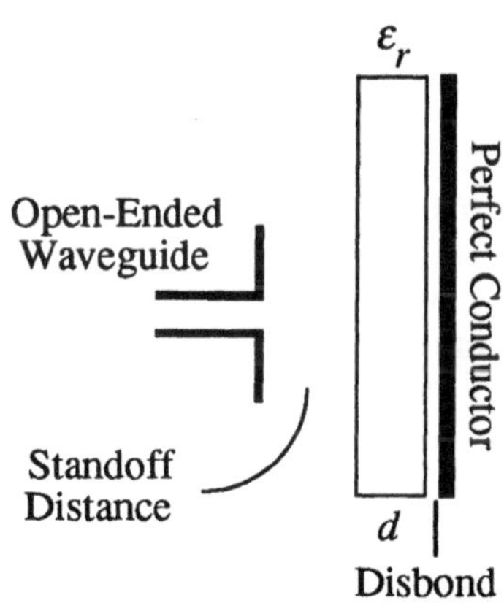

Fig. 3.24 Schematic of an open-ended rectangular waveguide, incorporating a standoff distance, radiating into a dielectric sheet with a disbond backed by a conducting plate.

This is intuitively due to the fact that when considering conductor backed cases, the signal travels through the medium twice, due to the total reflection of the signal at the conducting plate, and hence "sees" whatever dimension of interest twice.

It must be mentioned here that in practical applications the phase difference between a defective composite (e.g. with disbond) and that with no defect is usually desired. This is due to the fact that the phase of reflection coefficient is shown to be the more sensitive parameter. Thus, for instance in Fig. 3.26 the difference between these two cases for a disbond thickness range of 1 mm is about 150°. Since in this range the phase change is relatively linear, one may deduce a measured disbond detection sensitivity or resolution of better than 10 micrometers per degree. Hereon in this chapter, phase difference results will be primarily presented.

3.4.5 Measurement optimization of frequency and standoff distance[5]

In section 3.4.3 one example of frequency optimization for a dielectric coating backed by a conducting plate was demonstrated. In this section one of the most practically powerful aspects of this near-field microwave inspection technique; namely, the **measurement-parameter optimization** will be discussed for several typical cases. Given the geometry and composition of a stratified composite medium, the operating frequency and the standoff distance are the parameters that, when optimally chosen, can significantly increase measurement sensitivity to a parameter of interest such as disbond thickness or dielectric coating thickness. Another optimization tool, for composites backed by free-space, is the placement of a conducting plate a certain distance behind it, known as the back standoff distance or "backgap distance". This particular technique is shown to be very effective for measurement sensitivity enhancement when interested in detecting minute thickness variation in thin, low-permittivity and low-loss dielectric sheets. Clearly, when using this technique the optimal choice of the backgap distance becomes an important issue as well.

The geometry shown in Fig. 3.24 is considered first. To illustrate the frequency dependence of the detection sensitivity for a disbond between the dielectric sheet and the conducting plate, the following numerical simulation and measurements are carried out. The phase of reflection coefficient for a conductor backed lossy dielectric sheet with $\varepsilon_r = 9.55 - j1.1$ and a thickness of $d = 6.33$ mm at a standoff distance of 5.05 mm was calculated and measured for disbonds with thicknesses of 0, 0.8, 2 mm (disbond thickness equal to 0 mm is the no disbond case) at X-band, as

[5] Portions of this sub-section are reprinted with permission from (Transactions on Instrumentation and Measurement, IM-44, no. 2, pp. 326-28, April 1995) © 1995 IEEE, and with permission from (Research in Nondestructive Evaluation, Microwave Detection and Depth Determination of Disbonds in Low-Permittivity and Low-loss Thick Sandwich Composites, Qaddoumi, Zoughi and Carriveau, vol. 8, no. 1, pp. 51-63) © 1996 Springer-Verlag.

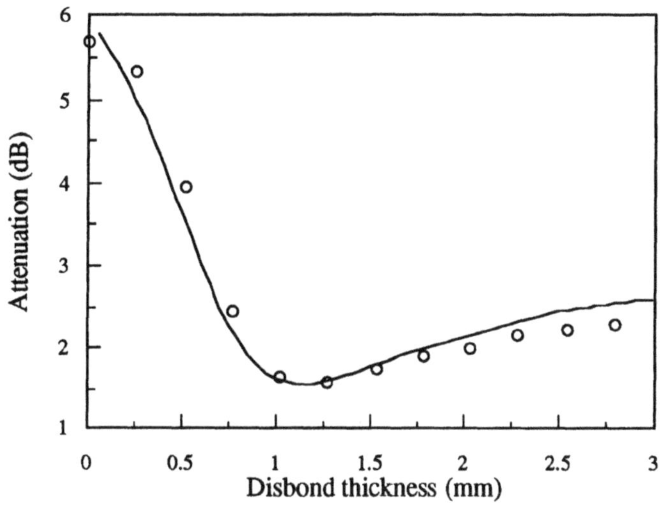

Fig. 3.25 Comparison of the calculated (solid line) and measured attenuation (circles) at 10 GHz, as a function of disbond thickness for the conductor backed composite shown in Fig. 3.24 (Bakhtiari *et al.*, 1994).

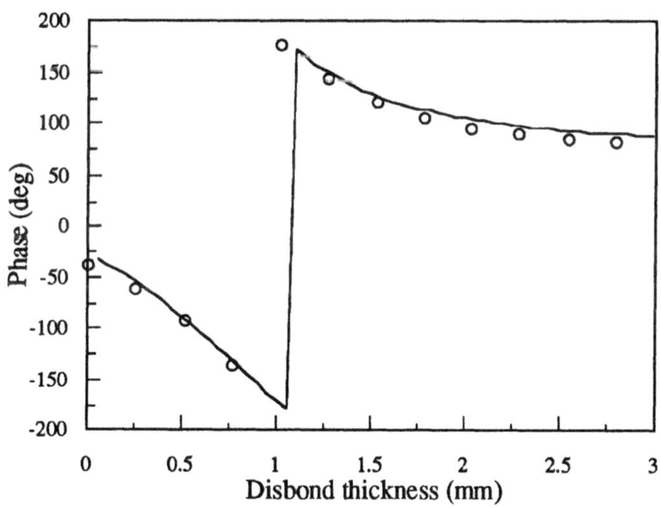

Fig. 3.26 Comparison of the calculated (solid line) and measured (circles) phase of reflection coefficient at 10 GHz, as a function of disbond thickness for the conductor backed composite shown in Fig. 3.24 (Bakhtiari *et al.*, 1994).

shown in Fig. 3.27 (Ganchev, Qaddoumi, Ranu *et al.*, 1995b). Clearly, there is good agreement between the calculated and the measured results. The deviation of measured results from the calculated results is attributed to slight uncertainties associated with setting the exact disbond thicknesses during the measurements as well as slight variations in the dielectric properties of the dielectric sheet with frequency. Next, the phase characteristics for each disbond are subtracted from the no disbond phase characteristic, and the resulting phase difference, as a function of frequency is shown in Fig. 3.28. The results demonstrate the measure of disbond detection sensitivity for this particular geometry and for this frequency range. For comparison purposes calculated results for smaller disbond thicknesses of 0.2 mm and 0.4 mm are also presented in Fig. 3.28. When the more industrially prevalent cases of small disbonds (0.2-0.8 mm) are considered, then the lower part of the frequency band (8.5-10 GHz) offers little detection sensitivity (to the disbond presence) compared to the upper portions of the frequency band (10.5-11.5 GHz). This example illustrates the importance of choosing the optimal operating frequency for these types of measurements.

Depending on the measurement requirements, the disbond detection and its thickness estimation may need to be conducted independent of small variations in the prescribed thickness of the dielectric sheet, *d*. This may be achieved by optimizing the operating frequency. Thus, the influence of varying the dielectric sheet thickness, in Fig. 3.24, on the phase difference is investigated at 10 GHz. Figure 3.29 shows the calculated phase difference between several disbond thicknesses and the zero disbond case as a function of the thickness of the dielectric sheet with $\varepsilon_r = 9.55 - j1.1$ and at a fixed standoff distance of 5.05 mm (Ganchev, Qaddoumi, Ranu *et al.*, 1995b). At this frequency and standoff distance, there is maximum sensitivity to disbond presence in the thickness range of 7-8 mm and to a lesser extent in the range of 12-13 mm. It must be noted that for these intervals, despite the attractive sensitivity, a small change in the dielectric sheet thickness, *d*, does not yield in an accurate estimation of the disbond thickness. Conversely, in the thickness intervals of 4-6 mm and 8.5-11 mm, the disbond detection and its thickness estimation are nearly independent of changes in the dielectric sheet thickness. For these ranges the disbond detection sensitivity is less; however, it may be adequate for many practical applications. To experimentally verify these results several measurements with different dielectric sheet thicknesses of 5.2 mm, 6.33 mm, 8.8 mm, and 10.1 mm were conducted at 10 GHz, as shown in Fig. 3.30. Again there is good agreement between the measured (discrete points) and the calculated (lines) results. If these lines are approximated to be straight (not true for all cases), their slope is then the measure of sensitivity to disbond detection. For example for d = 6.33 mm a sensitivity

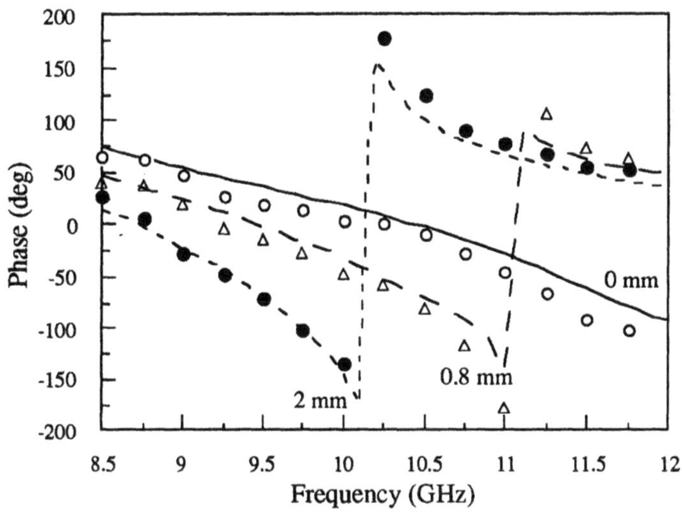

Fig 3.27 Comparison of the calculated (lines) and measured (discrete points) phase of reflection coefficient as a function of frequency for three different disbond thicknesses (Ganchev *et al.*, 1995b).

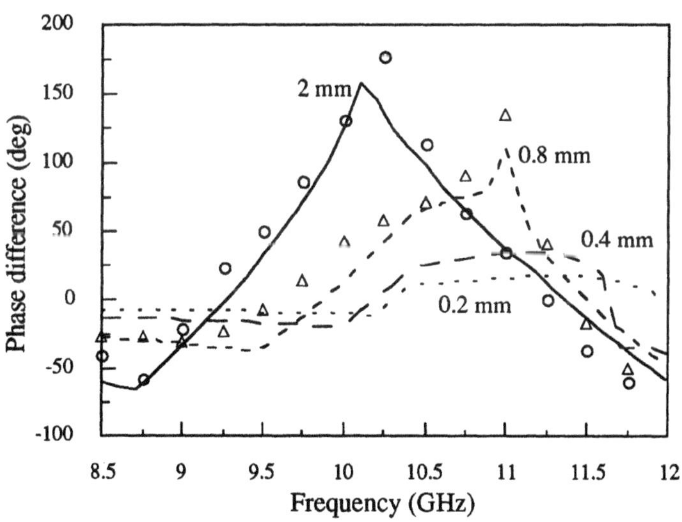

Fig 3.28 Calculated (lines) and measured (discrete points) phase differences for the results shown in Fig. 3.27 with calculated results for two other disbonds as well (Ganchev *et al.*, 1995b).

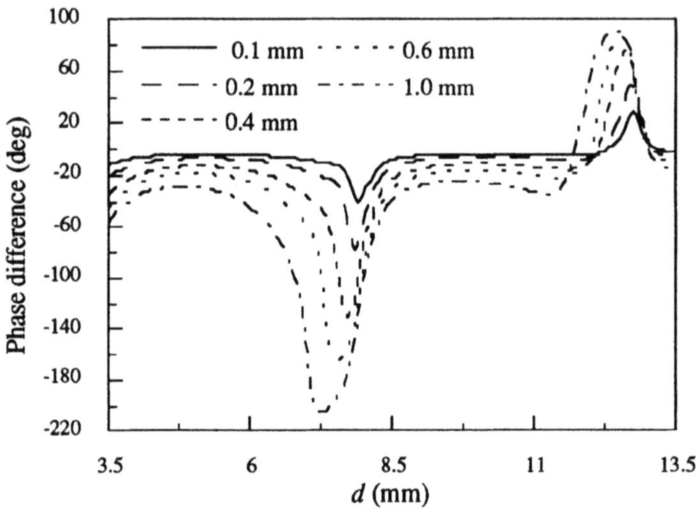

Fig. 3.29 Calculated phase difference as a function of dielectric sheet thickness for several disbond thicknesses at a frequency of 10 GHz (Ganchev *et al.*, 1995b).

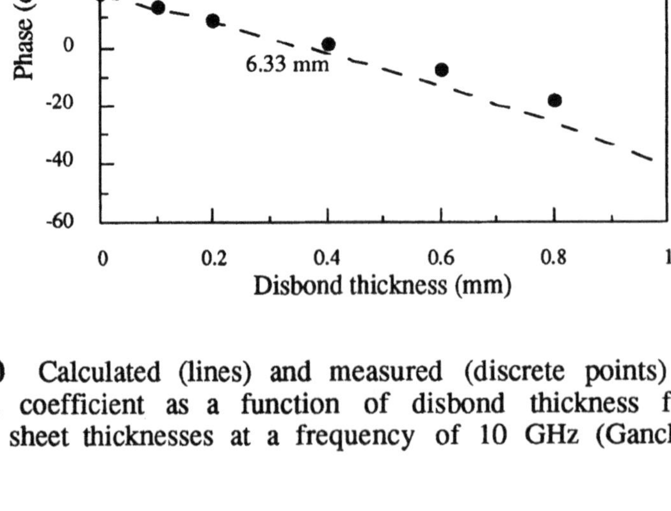

Fig. 3.30 Calculated (lines) and measured (discrete points) phase of reflection coefficient as a function of disbond thickness for several dielectric sheet thicknesses at a frequency of 10 GHz (Ganchev *et al.*, 1995b).

of 2.8° per 50 micrometer is obtained (for the given range of disbond thicknesses). Such sensitivity is sufficient in many production environments. Usually higher sensitivity requirements are accompanied with production of dielectric coatings with tight thickness tolerances, and in those cases the thicknesses around maximum detection sensitivity intervals (Fig. 3.29) may be used.

The geometry shown in Fig. 3.31 which constitutes a four-layer composite backed by a conducting plate is considered next. This geometry depicts layered composites in which a disbond may occur in between two dielectric sheets (Ganchev, Qaddoumi, Ranu *et al.*, 1995b).

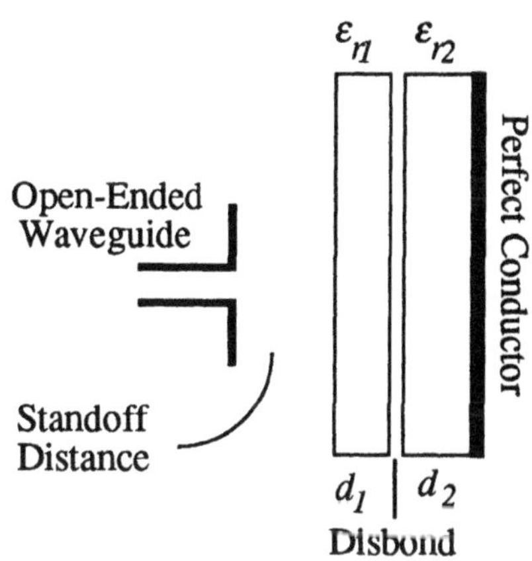

Fig. 3.31 Schematic of an open-ended rectangular waveguide, incorporating a standoff distance, radiating into a three-layer dielectric composite with a disbond in between the two layers, backed by a conducting plate.

Figure 3.32 shows the phase difference as a function of frequency for a fixed disbond thickness of 0.03 mm and a standoff distance of 5.05 mm for changing d_1 and d_2 while keeping $(d_1 + d_2)$ at a constant 8 mm. The dielectric sheets are assumed to have $\varepsilon_r = 9.55 - j1.1$. This case simulates a disbond whose location within a dielectric sheet varies. For this arrangement some frequencies, such as 9.75 GHz, are more sensitive to detecting the 0.03 mm-propagating disbond than other frequencies. Conversely, it is seen that at some depths from the surface, the disbond is more difficult to be detected (i.e. d_1 = 6 mm and d_2 = 2 mm). If the thickness of one of the dielectric layers is known (which may be the case in practice) the measurement may be optimized to be very sensitive to the presence of disbonds at fixed depths.

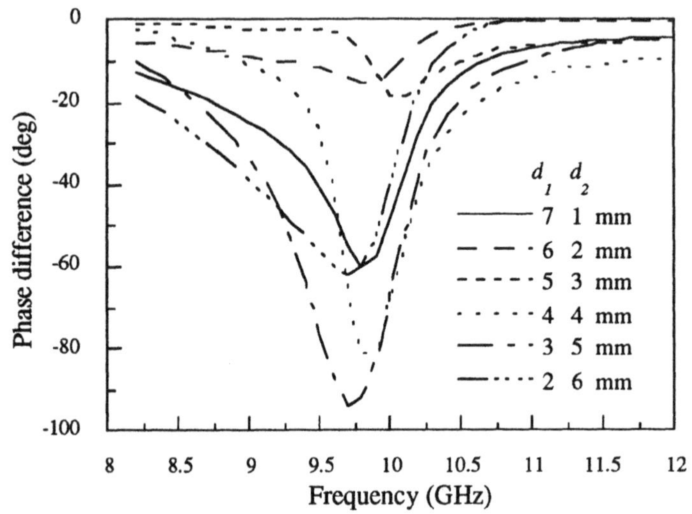

Fig. 3.32 Calculated phase difference as a function of frequency when a constant thickness disbond is at difference depths from a dielectric sheet front surface (Ganchev *et al.*, 1995b).

Next, the geometry depicted in Fig. 3.21 which constitutes a three-layer dielectric composite backed by free-space, while incorporating a standoff distance, is considered. For this cases, the calculated and measured phase difference for a standoff distance of 2.5 mm, dielectric sheet thicknesses of d_1 = 5.2 mm and d_2 = 6.33 mm (with the same dielectric properties as in the previous cases), and a varying disbond thickness at a frequency of 9.5 GHz are presented in Fig. 3.33. In this case the sensitivity to disbond thickness (calculated from the slope) is 1.3° per each 0.1 mm. the results thus far show that generally, better disbond detection sensitivity is achieved for the conductor backed case. In addition, the calculated results at 8.2 GHz are also presented which indicate that the sensitivity to disbond thickness is frequency dependent (Ganchev, Qaddoumi, Ranu *et al.*, 1995b).

The composite structures consider thus far have all been composed of two or less number of dielectric sheets containing disbonds. However, the theoretical derivation described earlier is capable of considering composite structures composed of many more layers. Such structures are commonly referred to as "thick sandwich composites". To show the applicability of this technique for inspecting such structures a thick sandwich composite was considered.

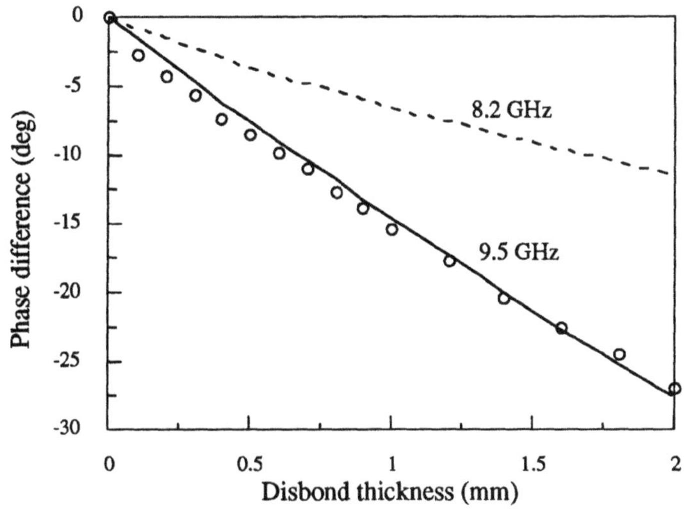

Fig. 3.33 Calculated and measured phase difference as a function of disbond thickness, for the composite shown in Fig. 3.21, at 9.5 GHz along with calculated results at 8.2 GHz (Ganchev *et al.*, 1995b).

The sandwich composite under consideration here, which could be considered as a representative example of a complex multi-layered composite structure, is composed of thirteen layers of dielectric materials, a standoff distance, and a free-space backing, as shown in Fig. 3.34. The layer-by-layer characteristics (thickness and the dielectric properties, ε_r) of this structure are as follows (Qaddoumi, Zoughi and Carriveau, 1996):

1. standoff distance - variable size,
2. skin laminate - 2.54 mm, ε_r = 4.5 - j0.045,
3. adhesive - 0.28 mm, ε_r = 3.1 - j0.01,
4. outer core - 45 mm, ε_r = 1.1 - j0.0026,
5. adhesive - 0.28 mm, ε_r = 3.1 - j0.01,
6. substrate - 0.14 mm, ε_r = 4.5 - j0.045,
7. adhesive - 0.28 mm, ε_r = 3.1 - j0.01,
8. inner core - 40.64 mm, ε_r = 1.1 - j0.0026,
9. adhesive - 0.28 mm, ε_r = 3.1 - j0.01,
10. substrate - 0.14 mm, ε_r = 4.5 - j0.045,
11. adhesive - 0.28 mm, ε_r = 3.1 - j0.01,
12. outer core - 45 mm, ε_r = 1.1 - j0.0026,
13. adhesive - 0.28 mm, ε_r = 3.1 - j0.01,
14. skin laminate - 2.54 mm, ε_r = 4.5 - j0.045.

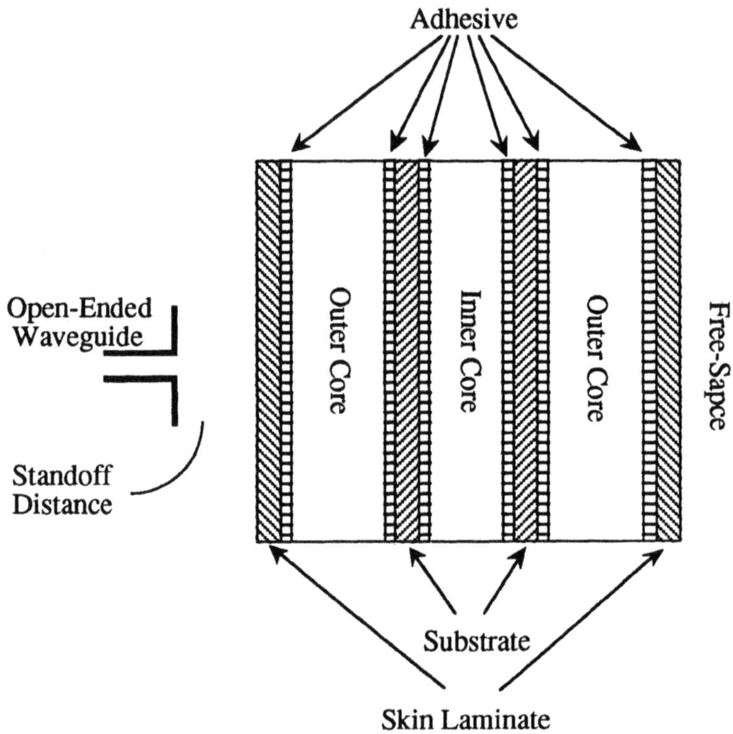

Fig. 3.34 Geometry of the thick sandwich composite structure, not-to-scale (Qaddoumi, Zoughi and Carriveau, 1996).

As shown, all layers are in the family of low-permittivity and low-loss dielectric materials, and the total thickness of the structure is approximately 14 cm. Since all layers are low-permittivity and low-loss dielectric materials, their dielectric properties are expected to remain relatively constant as a function of frequency. For all cases described here a disbond is assumed to replace an adhesive layer (disbond thickness is 0.28 mm). Therefore, there are six possible disbond locations in this structure. To obtain optimum measurement parameters for each disbond, the standoff distance was varied between 0 mm (in-contact) to 6 mm at one millimeter intervals. Several results at Ku-band (12-18 GHz), K-band (18-26.5 GHz) and Ka-band (26.5-40 GHz) frequency ranges are presented here (Qaddoumi, Zoughi and Carriveau, 1996).

(a) Standoff distance and frequency analyses

The phase of the reflection coefficient was calculated for this structure with and without a disbond (at all possible locations) at each standoff distance as a function of frequency. Subsequently, the phase difference between the no disbond and the disbonded cases was calculated and used to indicate the presence of the disbond. In the following discussions, the first, second, etc., disbond refers to the disbond between the skin laminate and the outer core, the outer core and the substrate, etc.

The phase difference at each standoff distance as a function of frequency (for Ku-band) was calculated for the first disbond (except at a 1-mm standoff due to some calculation difficulties) as shown in Fig. 3.35. The results indicate that a maximum phase difference of $\approx 9^{o}$ occurs at 18 GHz and a standoff distance of 3 mm. For this standoff distance the variation in the phase difference is relatively constant as a function of frequency. Relatively constant phase variation is a desirable feature from a practical point of view since no frequency selectivity is required, nor is there a tight requirement on keeping the frequency of operation constant. Collectively, the results shown in Fig. 3.34 are encouraging since, not only is there a great deal of frequency independence, but the phase difference due to the first disbond is adequately large for detection.

The phase difference as a function of frequency at all standoff distances for all other possible disbond locations was also calculated. These results were not as encouraging as those shown in Fig. 3.34 since the calculated phase difference for all of those cases was less than one degree (not adequate for detection). Therefore, one may conclude that this frequency band may not be optimum for detecting disbonds at all possible locations using standoff distances in the 0-6 mm range (Qaddoumi, Zoughi and Carriveau, 1996).

For K-band frequency range the 1-mm standoff distance produced the best results, particularly in the 23-26.5 GHz frequency range. Thus, only the results in this frequency range and at a standoff distance of 1 mm are shown here. Figure 3.36 shows the phase difference for the first disbond, while Fig. 3.37 shows the phase differences for all other disbonds, respectively. For the first disbond there is a large phase difference which is more than sufficient for its detection ($>120^{o}$ for a relatively wide range of frequencies). For the other disbonds the phase difference is sufficiently large for detection purposes (3°-8°). Furthermore, the magnitude and sign of the phase difference may be used for determining the location of a disbond, for example at a frequency of 22.5 GHz (Qaddoumi, Zoughi and Carriveau, 1996).

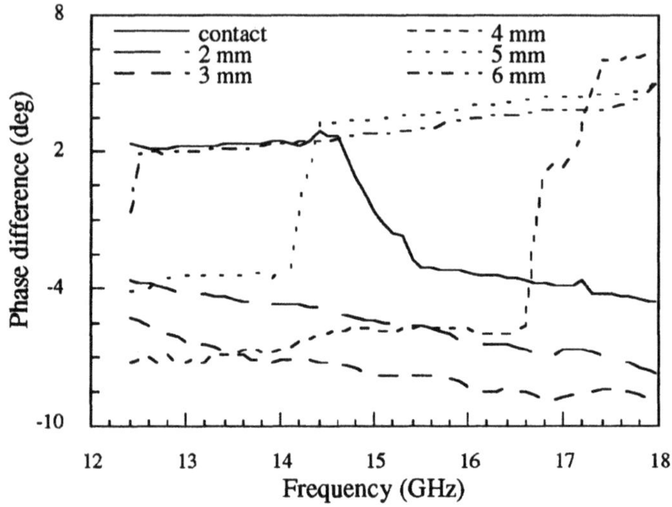

Fig. 3.35 Phase difference between no disbond and first disbond cases as a function of frequency (in Ku-band region), and at various standoff distances (Qaddoumi, Zoughi and Carriveau, 1996).

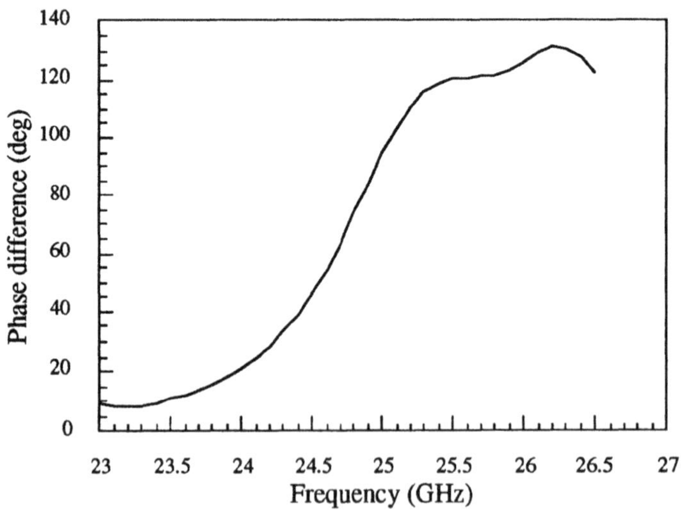

Fig. 3.36 Phase difference between no disbond and first disbond cases as a function of frequency (in K-band region), and at a standoff distance of 1 mm (Qaddoumi, Zoughi and Carriveau, 1996).

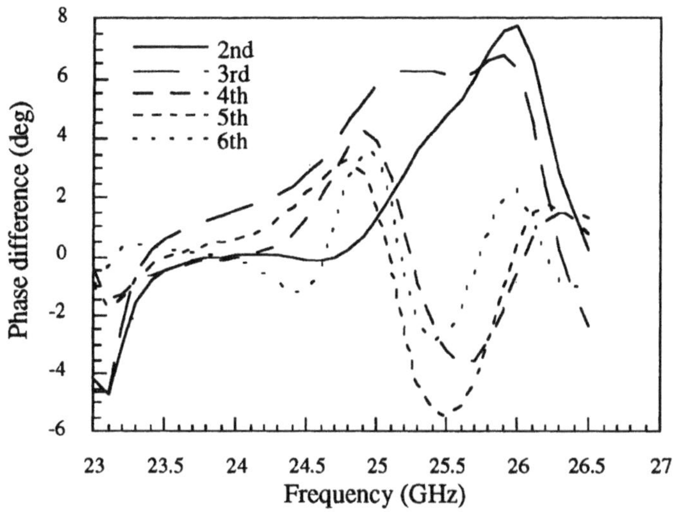

Fig. 3.37 Phase difference between no disbond and all other disbonds (except the first disbond) as a function of frequency (in K-band region), and at a standoff distance of 1 mm (Qaddoumi, Zoughi and Carriveau, 1996).

For Ka-band the phase difference for the first disbond at all standoff distances is shown in Fig. 3.38. It is clear that the phase difference associated with this disbond is quite large (e.g. 50° at a standoff distance of 2 mm and at a frequency of 33 GHz). Furthermore, there are several regions in which the phase difference is not only large but fairly constant (e.g., at a 1-mm standoff distance between 34 and 40 GHz). Figure 3.39 shows an example of the optimization process for the detection of the second disbond. The maximum phase difference is calculated to be $\approx 8^{\circ}$ at a combination of a 2-mm standoff distance and in the frequency range of 29-30.5 GHz. The maximum phase difference at any other combination does not exceed 2°. Thus, this disbond can be detected only with the proper choice of both of these parameters. Two issues are worth mentioning here. First, at a standoff distance of 2 mm the disbond is detected within a frequency band of $\approx$0.5 GHz which for some oscillators (e.g. cavity-tuned Gunn oscillators), may be wide enough not to cause concern about frequency drifts. Second, the results indicate that the standoff distance must be tightly held at 2 mm. This is an important practical issue since one may be required to use some type of a mechanical or electrical apparatus to keep the standoff distance constant. However, it is also possible (as shown earlier) to optimize the measurement parameters such that slight variations in the standoff distance do not cause the phase to change significantly.

A similar optimization process was performed for each disbond. A standoff distance of 2 mm was found to be optimal for detecting all disbonds at Ka-band. It has been shown that when operating at a combination of 2-mm standoff distance and in the frequency range of 29-29.5 GHz the detection of all disbonds may be possible (Qaddoumi, Zoughi and Carriveau, 1996).

(b) Potential of disbond location determination

This near-field microwave non-destructive testing technique has also shown the potential to determine the location (depth from surface) of a disbond in sandwich composites. In a sandwich composite of the type discussed here, there are only six discrete locations at which a disbond may be present. In the previous section it was shown that there are two parameters that may be used for uniquely determining the depth of a disbond; namely, the standoff distance and the operating frequency. Thus far, it has been established that disbonds, simulated by replacing an adhesive layer by free-space, can be detected using an optimal combination of the standoff distance and the operating frequency. Depth determination may be achieved by using several frequencies at a given standoff distance, several standoff distances at a given frequency, or a combination of both.

In all cases the first disbond can be unambiguously located (as shown earlier). Figure 3.40 (Ka-band) shows that the second and the third disbonds cluster together, and can be detected (not uniquely) at 29 GHz. Fourth, fifth and sixth disbonds also cluster together. The phase difference between the two clusters is $\approx 2.5^{\circ}$, which may be sufficient to differentiate between the two clusters.

At K-band (Fig. 3.37) a standoff distance of 1 mm was found to provide maximum sensitivity to the presence of disbonds in the frequency range of 23-26.5 GHz. Figure 3.36 shows that the sign of the phase difference, at 25.5 GHz, clusters the disbonds into two groups. The second and the third disbond have positive phase differences, but they cannot be uniquely distinguished. The fourth, the fifth and the sixth disbonds have negative phase differences, and only the fifth disbond can be uniquely distinguished. To resolve the ambiguity associated with identifying the location of the disbonds a second frequency of 25.3 GHz may be used. The phase differences at this frequency are listed in Table 3.4. Again, it is clear that first disbond can be unambiguously located. Furthermore, the second, the third and the fifth disbonds can be uniquely distinguished.

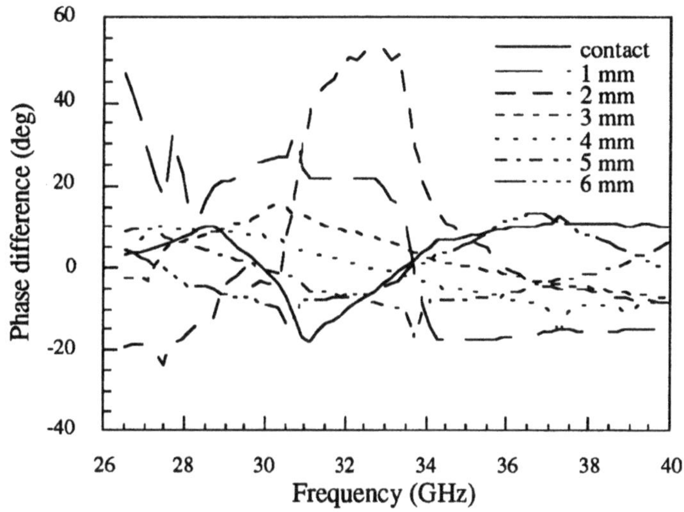

Fig. 3.38 Phase difference between no disbond and the first disbond as a function of frequency (in Ka-band region), and at various standoff distances (Qaddoumi, Zoughi and Carriveau, 1996).

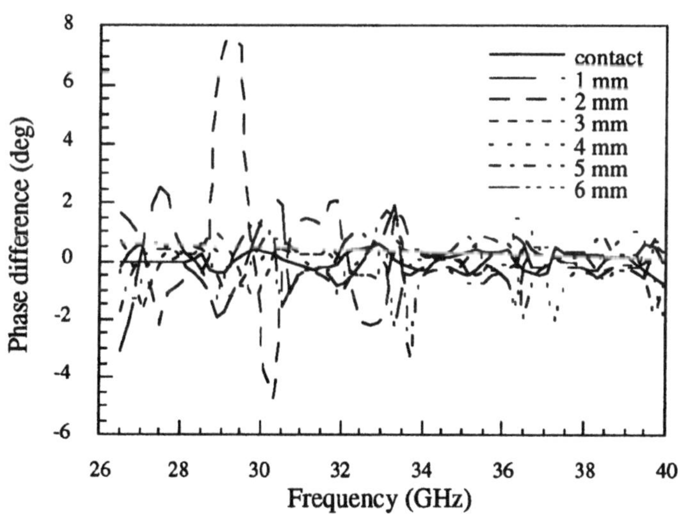

Fig. 3.39 Phase difference between no disbond and the second disbond as a function of frequency (in Ka-band region), and at various standoff distances (Qaddoumi, Zoughi and Carriveau, 1996).

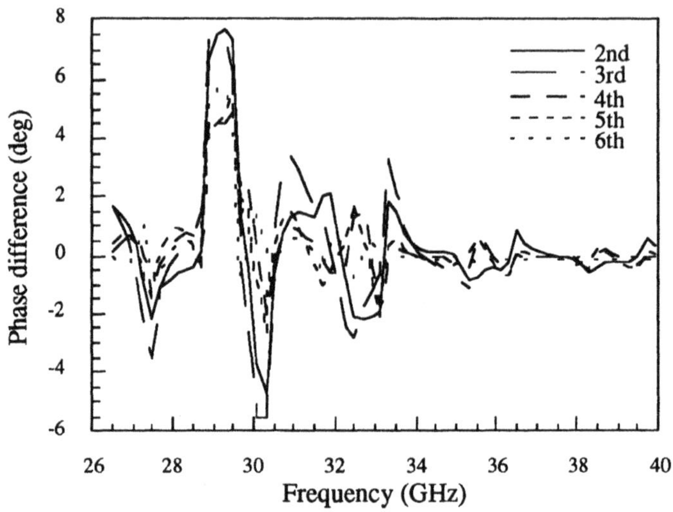

Fig. 3.40 Phase difference between no disbond and the second, third, fourth, fifth and sixth disbonds as a function of frequency (in Ka-band region), and at a standoff distance of 2 mm (Qaddoumi, Zoughi and Carriveau, 1996).

Finally, a third frequency of 25.7 GHz may be used to resolve the ambiguity associated with the fourth and the sixth disbonds (phase difference of $\approx 3^{\circ}$). Based on these results, one may conclude that with several sensors, each operating at a certain frequency and standoff distance unambiguous detection and location determination of all disbonds may be possible. Also, in some applications where there is access to both sides of the sandwich composite, the sixth, fifth, and fourth disbonds become the first, second, and third disbonds if inspected from the opposite side. These results indicate the potential and usefulness of standoff distance and frequency optimization when conducting near-field measurements using open-ended rectangular waveguides.

Table 3.4 Phase differences at 25.3 GHz and at 1 mm standoff distance for all disbonds (Qaddoumi, Zoughi and Carriveau, 1996).

1st	*2nd*	*3rd*	*4th*	*5th*	*6th*
115.8°	3.6°	6.3°	0.9°	-2.6°	-0.5°

3.4.6 Detection of rust under dielectric coatings

Detection and evaluation of rust or corrosion under paint and other dielectric coatings is an important issue is many industrial and military applications. Several investigations have been conducted on determining the capabilities of microwave techniques for this purpose (Qaddoumi, Shroyer and Zoughi, 1997; Qaddoumi, Handjojo, Bigelow *et al.*, 1999). When using this microwave technique to look for the presence of rust, the probing waveguide is considered to be radiating into a three layer conductor backed composite material. The first layer is the standoff distance, the second is the paint and the third is the layer of rust. Consider a case in which the paint thickness is 0.25 mm and the rust thickness is 0.1 mm at a frequency of 26 GHz. Using the measured dielectric properties of paint, Fe_2O_3 powder and black rust (from Table 2.10) Fig. 3.41 shows the phase difference between when there is rust (represented by Fe_2O_3 powder) and when there is no rust as a function of standoff distance. The results show that detection is a function of standoff distance and maximum detection occurs at a standoff distance of 1.5 mm. As mentioned in Chapter 2, real rust has higher dielectric properties than Fe_2O_3 powder. Therefore, similar results, when using black rust, are also shown in Fig. 3.41. In this case, the sensitivity to the presence of rust is increased by a factor of almost two. Moreover, in the standoff distance range of 2.5-4 mm the phase difference is relatively constant. This means that in this range there is no need to keep this parameter very constant.

3.4.7 Influence of backgap distance on measurement sensitivity[6]

Another parameter which may also be utilized for optimization is the backgap distance. The backgap distance is artificially created by placing a conducting plate behind a dielectric sheet or a multi-layered medium that contains a disbond. In this way, one may simultaneously optimize three measurement parameters; namely, the standoff distance, the frequency of operation and the backgap distance, resulting in increased measurement sensitivity levels (Gray and Zoughi, 1997).

Figures 3.42 and 3.43 show two specific geometries of interest; namely, a dielectric sheet, and a disbond between two dielectric sheets, respectively. In both figures a conducting plate is placed behind the medium under inspection providing for the backgap distance. In either case, the incident

---

[6] Portions of this sub-section are reprinted with permission from *Materials Evaluation*, vol. 55, no. 1, pp. 42-48, Copyright 1997 © The American Society for Nondestructive Testing, Inc.

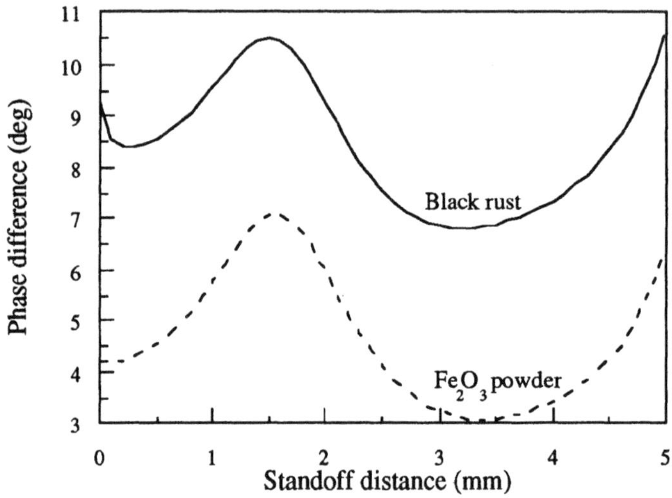

Fig. 3.41 Phase difference between rust and no rust (Fe_2O_3 powder and black rust) cases for a paint thickness of 0.25 mm and a rust thickness of 0.1 mm at a frequency of 26 GHz.

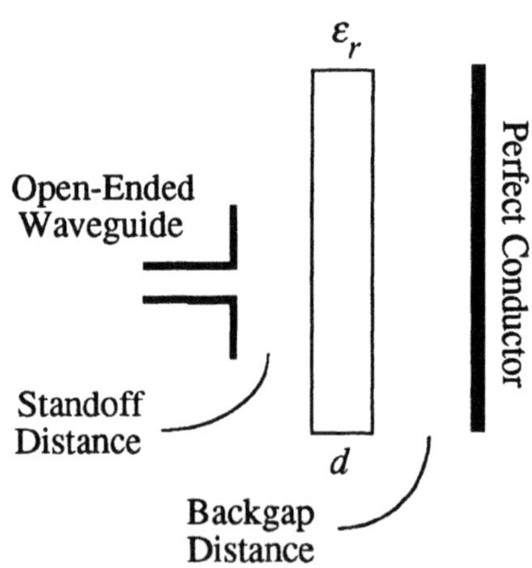

Fig 3.42 An open-ended rectangular waveguide radiating into a single dielectric sheet while incorporating a standoff distance and a backgap distance in the measurement scheme (Gray and Zoughi, 1997).

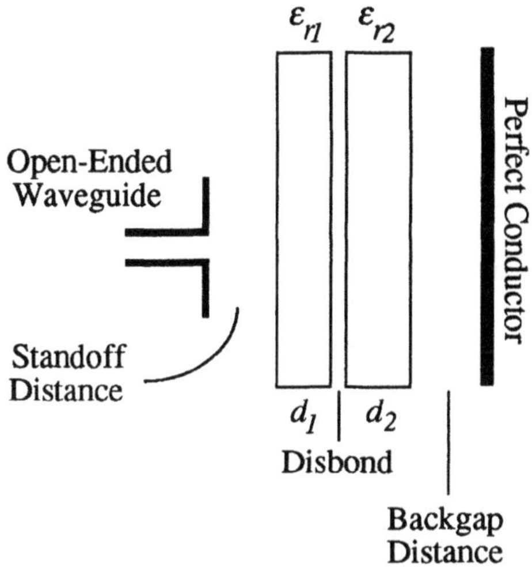

Fig 3.43 An open-ended rectangular waveguide radiating into a two layer composite medium with a disbond while incorporating a standoff distance and a backgap distance in the measurement scheme (Gray and Zoughi, 1997).

signal travels twice through the medium under test, hence rendering a more sensitive measurement.

As mentioned earlier, the goal of incorporating a backgap distance is to show that by optimally choosing the standoff distance, the frequency of operation and the backgap distance, higher measurement sensitivities may be achieved than when no backgap distance is used (i.e. specimen under inspection backed by free-space). The results of the following cases are presented in this section:

- thin dielectric sheet thickness variation (refer to Fig. 3.42) with a nominal sheet thickness of d = 1 mm and a thickness variation of Δd = 0.05 mm (50 micrometers), at X-band (8.2-12.4 GHz) and Ka-band (26.5-40 GHz),
- thick dielectric sheet thickness variation with a nominal sheet thickness of d = 5 mm and a thickness variation of Δd = 0.1 mm (100 micrometers), at X-band (8.2-12.4 GHz) and K-band (18-26.5 GHz) and,
- disbonds (refer to Fig. 3.43) with thicknesses of 0.05 and 0.1 mm at X- and K-bands when the nominal thickness of the layered composite is 5 mm composed of two layers with the same ε_r = 9.55 - j1.1, and with individual thicknesses of d_1 = 2 mm and d_2 = 3 mm.

it must be mentioned that this approach applies to more complicated layered composite structures as well.

For the infinite half-space backed case, theoretical optimization is achieved by fixing one of the measurement parameters such as the frequency of operation and then changing the standoff distance until a maximum phase difference, between the nominal thickness d and $(d + \Delta d)$ is obtained. Then, at this standoff distance, the frequency (within the given band) is changed to investigate any increases in the phase difference. This iterative process is continued until the optimal standoff distance and operating frequency are obtained. The same approach is followed for all other cases.

(a) Thin dielectric sheet thickness variation (X-band and Ka-band)

The theoretical approach used for the optimization technique (for the free-space backed case) is initiated by arbitrarily choosing the operating frequency of 10 GHz. The standoff distance is then varied between 0 and 5 mm for both thicknesses of 1 mm and 1.05 mm. The maximum phase difference is calculated to be 5.3° at a standoff distance of 5 mm. Next, the standoff distance of 5 mm is used while the frequency is varied in the X-band frequency range (8.2-12.4 GHz). Figure 3.44 shows the results of phase difference for this calculation. The results indicate that a maximum phase difference of 5.3° occurs at an operating frequency of 9.75 GHz. Next, the operating frequency of 9.75 GHz is used and the standoff distance is varied again between 0 and 5 mm. This last iteration yields the optimal values for the standoff distance and the operating frequency to be 5 mm and 9.75 GHz, respectively, resulting in a detectable maximum phase difference of 5.3° for a thickness variation of 0.05 mm. Once again it is important to note that this level of phase sensitivity is achieved at a wavelength of approximately 30.77 mm (in free-space). The results also show that if there is drift in the optimal frequency of 9.75 GHz within a few hundred MHz (in this case ≈9.2-10.3 GHz), the phase difference remains relatively unchanged (≈5°). Thus once again, problems associated with oscillator frequency stability are minimized, and a frequency drift of about 1 GHz may be tolerated.

Next, the optimization scheme incorporating a backgap distance (Fig. 3.42) is invoked by arbitrarily choosing an operating frequency in the X-band frequency range and a backgap distance, while calculating the best standoff distance, and continue this iterative process, as explained earlier until the maximum phase difference is obtained. Figure 3.45 shows that at an optimal standoff distance of 1.8 mm, an operating frequency of 10.25 GHz, and a backgap distance of 1.5 mm, a maximum phase difference of 104° is obtained. This is a significant improvement when compared with 5.3° which was obtained without incorporating the backgap distance.

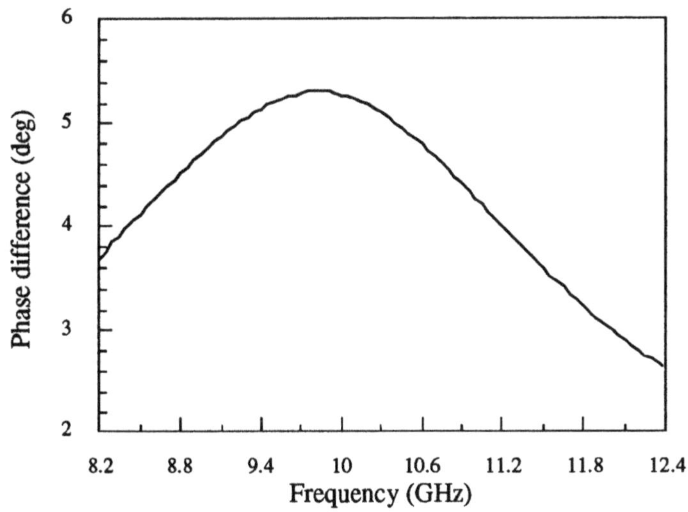

Fig 3.44 Phase difference between two thin dielectric sheets as a function of frequency (X-band) at a standoff distance of 5 mm (Gray and Zoughi, 1997).

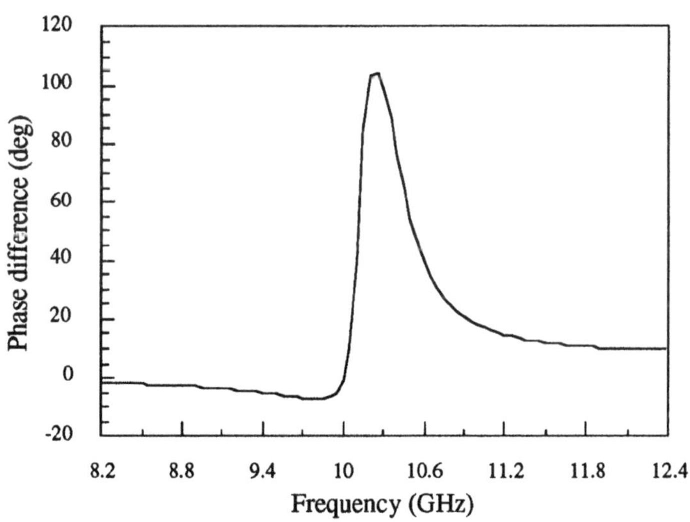

Fig 3.45 Phase difference between two thin dielectric sheets as a function of frequency (X-band) at a standoff distance of 1.8 mm and a backgap distance of 1.5 mm (Gray and Zoughi, 1997).

The phase difference between two sheets with different thicknesses is not expected to always be a linear function of the thickness difference (Δd). However, for thin sheets and small thickness variations, one may consider the phase difference to remain relatively linear as a function of (Δd). Under this assumption and using the results of Fig. 3.45, one may estimate the phase variation due to 0.01 mm (ten micrometers) of dielectric sheet variation is approximately 21°. This translates to a phase difference of about 2° per one micrometer of thickness variation. Considering the fact that the operating optimal wavelength is 30.77 mm (in free-space), the sensitivity to thickness variation for this case is indeed extremely high.

Figure 3.46 shows the optimal results, at Ka-band (26.5-40 GHz), with and without a backgap distance. Without the backgap distance, the maximum phase difference is 14.6° at a standoff distance of 4.68 mm and an operating frequency of 40 GHz. However, with the backgap distance, the maximum phase difference increases to 31.4° at a standoff distance of 5 mm, an operating frequency of 36.9 GHz and a backgap distance of 2.8 mm. Thus, adding a backgap distance to the system more than doubles the sensitivity to a 0.05 mm thickness increase in the dielectric sheet, however compared to the X-band results (Fig. 3.45) the calculated sensitivity is less. This observation clearly indicates that when operating in the near-field of an open-ended rectangular waveguide probe an increase in the operating frequency does not always render improved resolution/sensitivity along the direction of propagation. This may be partially due to the fact that the same dielectric properties were used in these calculations for the entire frequency band.

(b) Thick dielectric sheet thickness variation (X-band)

Figure 3.47 shows the optimal calculated results with and without a backgap distance in the X-band frequency range. Without a conductor backing a maximum phase difference of 147.1° is obtained for a standoff distance of 3.15 mm and a frequency of 9.5 GHz. Whereas, with a backgap distance of 4.85 mm a maximum phase difference of 163.9° is obtained at a standoff distance of 2 mm and a frequency of 10.15 GHz. The results show that at this frequency band although the addition of a backgap distance improves the calculated sensitivity, however this improvement is not as substantial as in the thin sheet case.

For the cases investigated here (i.e. lossy dielectric sheets), it is evident that the sensitivity to thickness variation is much higher for thin sheets than for thick sheets. For these cases this may be expected for the following reasons. Placing a conducting plate behind a dielectric sheet provides for the incident signal to travel through the sheets twice, once via the forward traveling wave and once via the reflected wave off of the conducting plate.

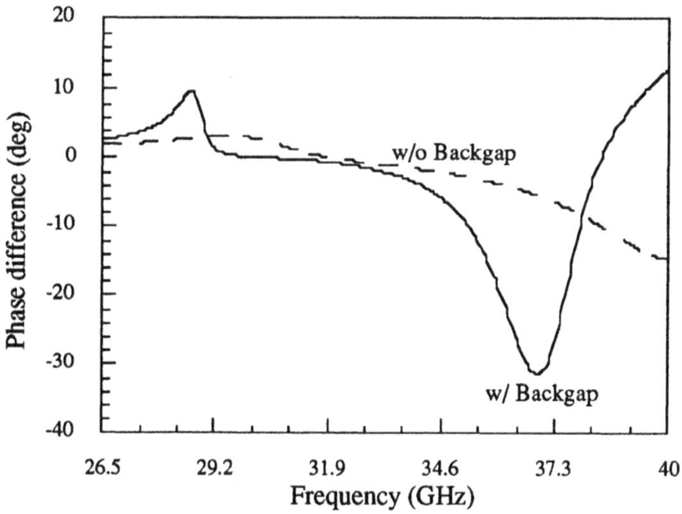

Fig 3.46 Phase difference between two thin dielectric sheets as a function of frequency (Ka-band) with and without a backgap distance (Gray and Zoughi, 1997).

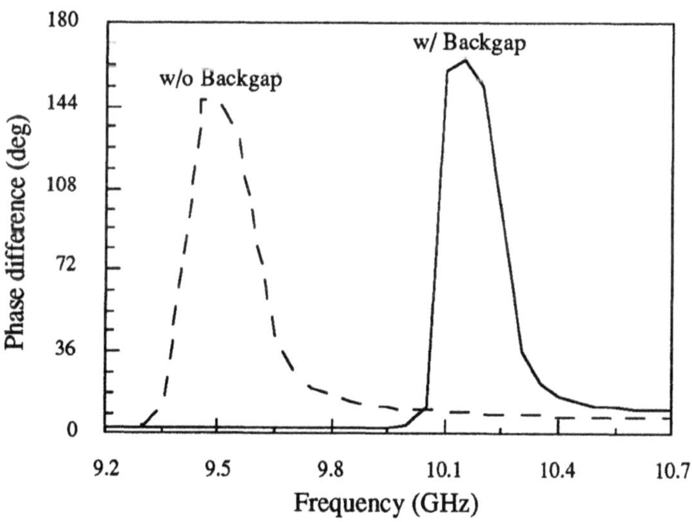

Fig 3.47 Phase difference between two thick dielectric sheets as a function of frequency (X-band) with and without a backgap distance (Gray and Zoughi, 1997).

Therefore, this may be interpreted as "the sheet having been inspected twice". As mentioned earlier, this fact is the primary reason why the incorporation of a conducting plate generally tends to increase the measurement sensitivity. However, when a sheet is thick and made of a lossy dielectric material (as in the cases described here), the incident signal attenuates significantly when traveling through the material. Whatever signal which passes through the sheet is reflected off of the conducting plate and travels through the sheet experiencing a similar attenuation level. Thus, this portion of the signal picked up by the open-ended waveguide may be relatively negligible so as not to provide much more information than when there is no conducting plate present. In addition, for thick and lossy dielectric sheets the far-field criterion is satisfied earlier in the dielectric sheet, rendering less phase sensitivity. One must be careful to consider the thickness of the sheet as a combination of its physical thickness as well as its dielectric properties (lossy vs. low-loss) and the frequency of operation (i.e. electrical thickness).

For all cases the optimal calculated parameters are dependent on the two geometries described here. The parameters calculated with a backgap distance will not necessarily work for the case without a backgap distance. To illustrate this, the frequency found with a backgap distance is used for the case without, and then the standoff distance is varied. Figure 3.48 shows that at the frequency determined with a backgap distance, the maximum phase difference for the case without a backgap distance is less than that calculated earlier. Next, the frequency determined for the case without a backgap distance is used for when there is a backgap distance and the standoff distance is changed. The optimal backgap distance of 4.85 mm determined earlier is used here. Figure 3.48 also shows that at the optimal frequency found for without a backgap distance and used along with a backgap distance of 4.85 mm results in a maximum phase difference which is less than that found earlier. This exercise shows that the optimization parameters are clearly dependent on the geometry of the system and a set of optimal parameters for a certain geometry may not be optimal for the other geometry.

(c) Disbond detection (X-band and K-band)

The geometry for this case is shown in Fig. 3.43. Disbonds with thicknesses of 0.05 and 0.1 mm are analyzed using both approaches and compared to the case with no disbond, respectively. Figure 3.49 shows the calculated phase as a function of the operating frequency without a backgap distance.

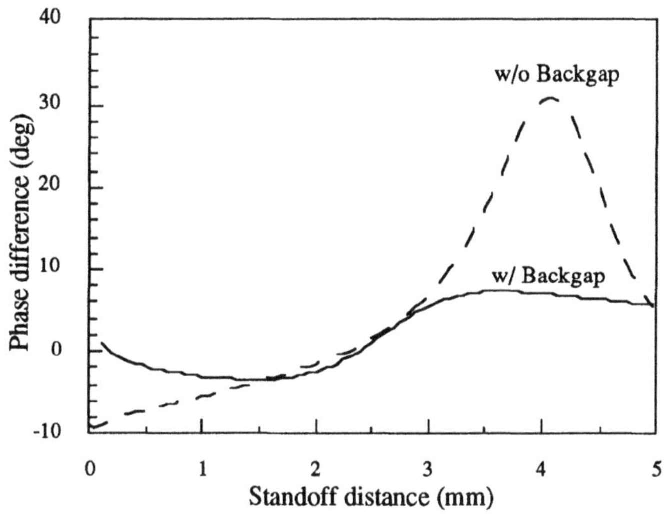

Fig 3.48 The effect of using optimal frequency for one case and using it to calculate the phase difference for the other case (Gray and Zoughi, 1997).

Figure 3.50 shows the phase difference as a function of the operating frequency for these cases. For a disbond thickness of 0.05 mm, the maximum phase difference is 68° at a standoff distance of 3.25 mm and an operating frequency of 9.55 GHz, whereas for the 0.1 mm-thick disbond the maximum phase difference is 124.5^o at a standoff distance of 3.25 mm and an operating frequency of 9.475 GHz. Note the frequency at which the phase reversal occurs for each graph in Fig. 3.49. For these three cases, one may use the frequency at which the phase reversals occur as an indication of the disbond thickness since this frequency also provides information about the disbond thickness for practical applications.

The results of phase difference as a function of the operating frequency with a backgap distance is shown in Fig. 3.51. For the 0.05 mm-thick disbond, the maximum phase difference is 120° at a standoff distance of 2 mm, at an operating frequency of 10.2 GHz, and a backgap distance of 5 mm. For the 0.1 mm-thick disbond, the maximum phase difference is 135° at the same standoff distance, operating frequency and backgap distance. These results show that for the thinner disbond an added backgap renders much better detection sensitivity than for the thicker disbond. This could be interpreted as saying that the introduction of a backgap distance provides for detecting disbonds just as soon as they have been initiated (i.e. very thin) in such a two-layer medium.

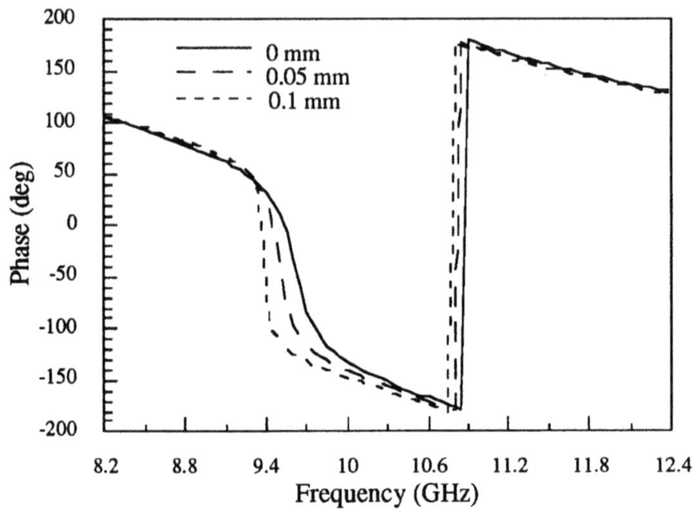

Fig 3.49 Phase of reflection coefficient as a function of frequency (X-band) for no disbond and two different disbond thicknesses without a backgap distance (Gray and Zoughi, 1997).

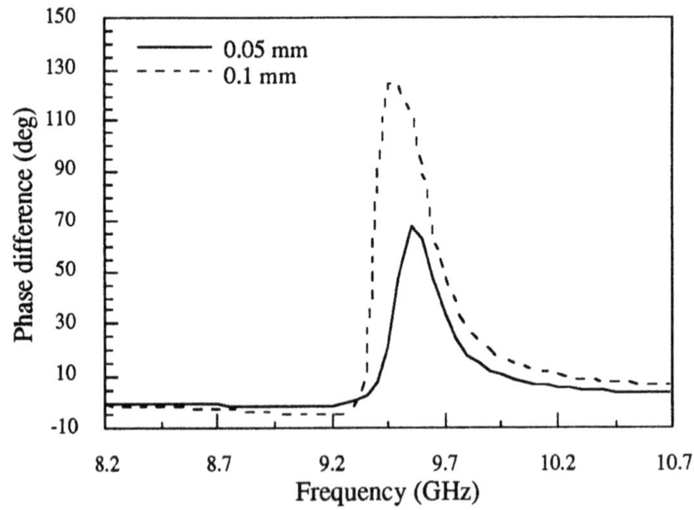

Fig 3.50 Calculated phase difference between the no disbond case and the two disbonded cases shown in Fig. 3.48 (Gray and Zoughi, 1997).

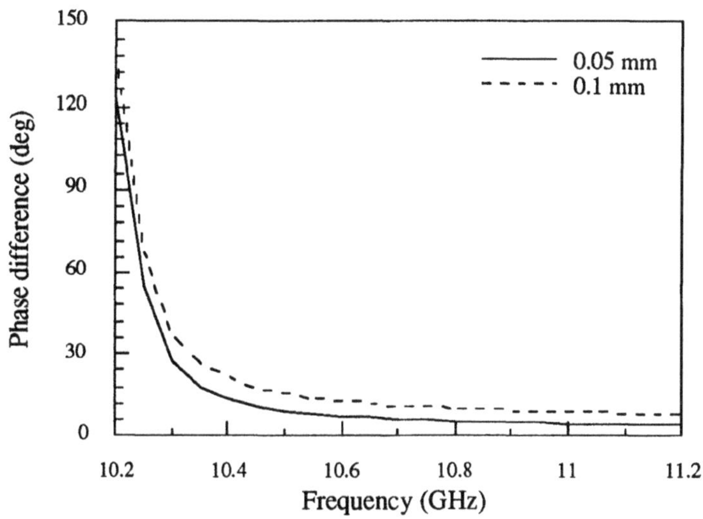

Fig 3.51 Calculated phase difference as a function of frequency (X-band) for no disbond and two disbond thicknesses with a backgap distance of 5 mm and a standoff distance of 2 mm (Gray and Zoughi, 1997).

Figure 3.52 shows the results of phase difference as a function of standoff distance without a backgap distance for K-band. For the 0.05 mm-thick disbond, the maximum phase difference is 24.9° at a standoff distance of 1.9 mm and an operating frequency of 19.9 GHz, whereas for the 0.1 mm-thick disbond the maximum phase difference is 39.3° at the same standoff distance and operating frequency. However, with a backgap distance (Fig. 3.53) and for the 0.05 mm-thick disbond, the maximum phase difference increases to 65.7° at a standoff distance of 2.2 mm, an operating frequency of 20.4 GHz and a backgap distance of 5 mm. Similarly, for the 0.1 mm-thick disbond the maximum phase difference increases to 90.1° at a standoff distance of 2.2 mm, an operating frequency of 20.4 GHz and a backgap distance of 4.9 mm. These results indicate that for this frequency band the addition of a backgap distance renders significantly higher detection sensitivity to the presence of a disbond.

(d) Measurement results

To illustrate the experimental utility of incorporating a backgap distance, a limited number of experiments were conducted on thin and low-loss dielectric (plastic) sheets. Two thin plastic sheets with thicknesses of 0.102 mm and 0.178 mm were chosen for an experiment at 38 GHz, demonstrating the experimental utility of this approach for dielectric sheet thickness variation detection. A custom designed phase detector microwave

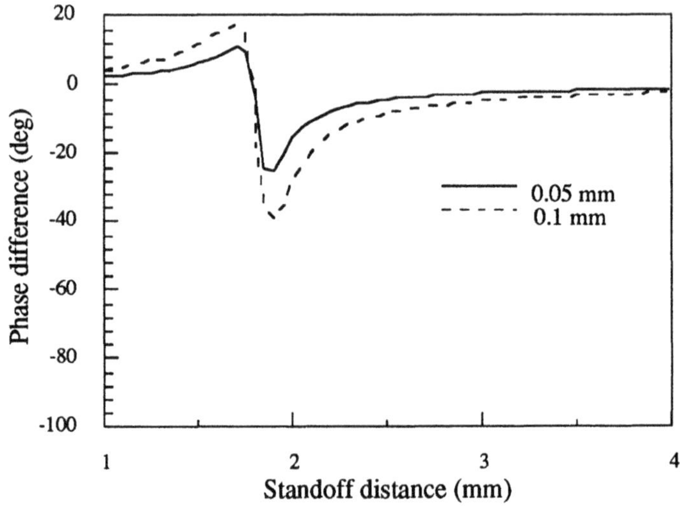

Fig 3.52 Phase difference, as a function of standoff distance, between the no disbond case and two disbond thicknesses without a backgap in K-band (Gray and Zoughi, 1997).

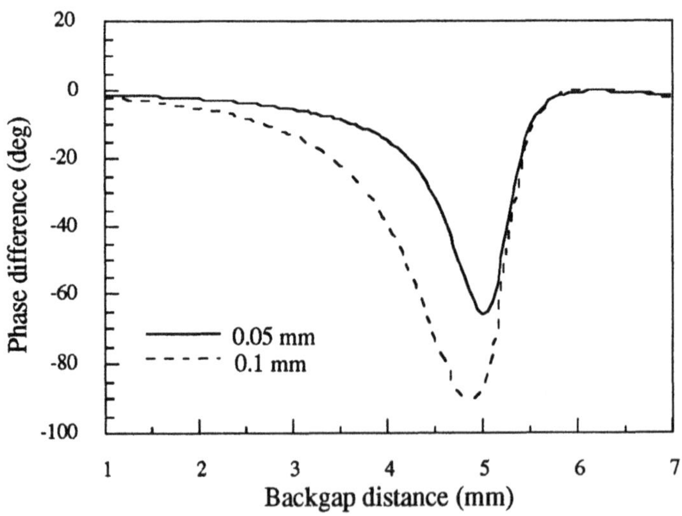

Fig 3.53 Phase difference, as a function of standoff distance, between the no disbond case and two disbond thicknesses with a backgap in K-band (Gray and Zoughi, 1997).

circuit was assembled employing a dual-directional coupler and a mixer. The output of the mixer which is a dc voltage is proportional to the phase of the effective reflection coefficient. Using this setup, a case without a backgap distance is compared to two cases with two different backgap distances (the measurement parameters are not necessarily optimal for this situation, however the results indicate the utility of this technique). Figure 3.54 shows the voltage variation as a function of the standoff distance, without a backgap distance, for both plastic sheets. The maximum voltage difference between the two thickness is ≈5 mV at a standoff distance of 3.75 mm. For comparison, Fig. 3.55 shows the voltage difference as a function of standoff distance for the sheets when backgap distances of 6.4 mm and 12.7 mm are added to the measurement setup, respectively. When a backgap is not present the maximum measured voltage difference is ≈5 mV between the two thin plastic sheets. When backgap distances of 6.4 mm and 12.7 mm are present, the maximum measured voltage differences increase to 58 mV and 17 mV, respectively. The addition of the backgap distance has clearly increased the measurement sensitivity to the difference in thickness between these two thin plastic sheets (e.g. adding a backgap of 6.4 mm to the geometry of the system improves the measurement sensitivity by a factor of ≈12 times). Furthermore, the dependence of the measurement sensitivity to the backgap distance is also demonstrated.

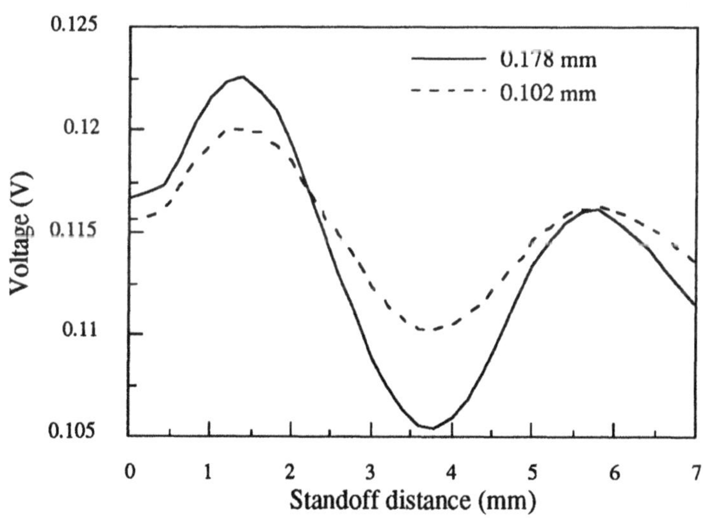

Fig 3.54 Measured voltage for two thin plastic sheets as a function of standoff distance without a backgap distance at a frequency of 38 GHz (Gray and Zoughi, 1997).

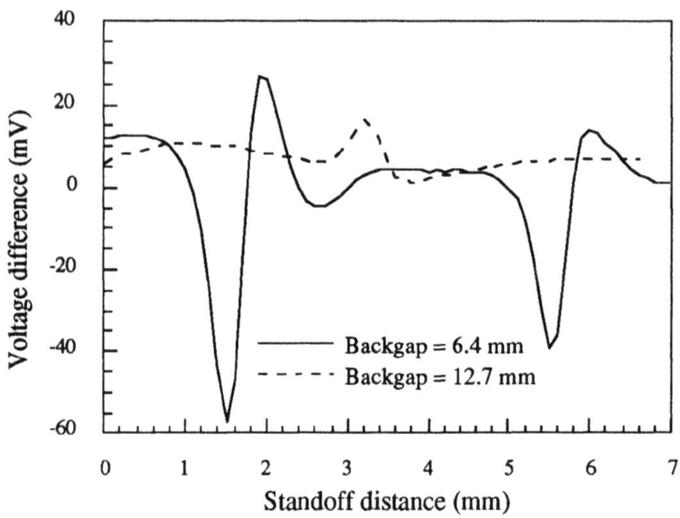

Fig 3.55 Measured voltage differences between two thin plastic sheets as a function of standoff distance for two different backgap distances at a frequency of 38 GHz (Gray and Zoughi, 1997).

3.5 SUMMARY

A theoretical model and its results along with some experimental results were presented in conjunction with near-field microwave inspection of layered dielectric composites using an open-ended rectangular waveguide. The theoretical analysis is based on a Fourier transform boundary matching technique to construct the field components in a layered media outside an open-ended rectangular waveguide coupled with the variational form of the terminating admittance of the flange-mounted aperture. The integrity of the calculated results were examined by conducting several measurements. Variations of attenuation and phase of reflection coefficient as a function of parameters such as the operating frequency, the standoff distance and dielectric sheet and disbond thicknesses were investigated. The experimental results showed good agreement with the calculated results. The results indicated that high resolutions may be achieved in examination of generally lossy dielectric sheets without the need to operate at very high microwave or millimeter wave frequencies. Furthermore, they demonstrate the sensitivity of such versatile technique as it applies to the inspection of multi-layered composite media, and the importance of a fast and reliable numerical model as a tool to gain real time *a priori* knowledge of the underlying process. It is concluded that the theoretical model can render important information for optimization of such measurement parameters as the standoff distance and operating frequency leading to better resolutions.

The sensitivity of the disbond detection is shown to be dependent on the frequency of operation and the geometry of the layered dielectric composite. Depending on the measurement requirements, the disbond detection and the estimation of its thickness may be independent of the changes in the dielectric coating thickness. The results of the theoretical analysis for detecting disbonds in a thick sandwich composite and unambiguously determining their locations were also presented. The results indicate that, for this specific type and family of sandwich composite structures, disbond detection at different depths is possible at a number of frequencies and standoff distances. It was shown that a combination of frequencies and standoff distances may be used for unambiguous depth determination. Several sensors, each operating at a different frequency and standoff distance may be used to unambiguously detect and locate disbonds. The disbond thickness used in this study was assumed to be equal to the thickness of the adhesive layer. In practice, however, a disbond thickness may be greater. In such cases the phase difference increases substantially, rendering the disbonds much easier to detect. Additionally, disbond depth determination in such cases may be easier as well. Since disbond thickness also influences the phase of the reflection coefficient, it is very likely that this technique can be used to determine not only its depth but also its thickness (within a given range) as well.

Finally, an extremely sensitive microwave approach for dielectric sheet thickness variation detection and disbond detection in a two-layer dielectric composite was described. This approach utilized a conducting plate behind the specimen under inspection providing for a backgap distance. This parameter along with the standoff distance and the frequency of operation may be collectively optimized to increase the sensitivity of detection to the parameters of interest (i.e. dielectric sheet thickness variation). The results of calculations show that in most cases the addition of a backgap distance significantly improves the sensitivity. However, this increase is dependent upon the electrical thickness of the medium under test which is a function of its thickness, dielectric properties and the frequency of operation. The utility of the electromagnetic model to predict such increase in the measurement sensitivity was effectively demonstrated. It was also shown that the influence of the backgap distance is more pronounced when applied to thinner dielectric sheets as opposed to thicker. This is due to the fact that the microwave signal traveling through the sheet (for the cases discussed in this chapter) is substantially attenuated which reduces the influence of the backgap distance. Experimental results conducted on thin plastic sheets also verified the utility of adding a backgap distance to the measurement scheme.

REFERENCES

Bahr, A.J. (1982) Microwave nondestructive testing methods, Gordon & Breach, Newark, NJ.

Bakhtiari, S. and Zoughi, R. (1990) Microwave thickness measurement of lossy layered dielectric slabs using incoherent reflectivity, *Research in Nondestructive Evaluation*, **2** (3), pp. 157-68.

Bakhtiari, S. (1992) Open-ended sensors for microwave nondestructive evaluation of layered composite media, Ph.D. Dissertation, Colorado State University.

Bakhtiari, S., Ganchev, S. and Zoughi, R. (1993a) Microwave swept-frequency optimization for accurate thickness or dielectric property monitoring of conductor-backed composites, *Materials Evaluation*, **51** (6), pp. 740-43.

Bakhtiari, S., Ganchev, S. and Zoughi, R. (1993b) Open-ended rectangular waveguide for nondestructive thickness measurement and variation detection of lossy dielectric slabs backed by a conducting plate, *IEEE Transactions on Instrumentation and Measurement*, **42** (1), pp. 19-24.

Bakhtiari, S., Qaddoumi, N., Ganchev, S. *et al.* (1994) Microwave non-contact examination of disbond and thickness variation in stratified composite media, *IEEE Transactions Microwave Theory and Techniques*, **42** (3), pp. 389-95.

Bakhtiari, S., Gopalsami, N. and Raptis, A.C. (1995), Characterization of delamination and disbonding in stratified dielectric composites by millimeter wave imaging, *Materials Evaluation*, **53** (4), pp. 468-71.

Bakhtiari, S., Ganchev, S. and Zoughi, R. (1995) A generalized formulation for admittance of an open-ended rectangular waveguide radiating into stratified dielectrics, *Research in Nondestructive Evaluation*, **7** (2/3), pp. 75-87.

Balanis, C.A. (1997) Antenna theory, analysis and design, Wiley & Sons, New York.

Boving, K.G. (1989) NDE handbook, non-destructive examination methods for condition monitoring, *Teknisk Forlag A/S*, Danish Technical Press.

Carriveau, G.W. (1994) Technology assessment on nondestructive evaluation applied to thick sandwich composites, *NTIAC Report A7404-300:GWC-D164.3.*

Collin, R.E. (1960) Field theory of guided waves, McGraw-Hill, New York.

Compton, R.T. (1964) Admittance of aperture antennas radiating into lossy media, Ph.D. Dissertation, Ohio State University.

Croswell, W.F., Rudduck, R.G. and Hatcher, D.M. (1967) Admittance of a rectangular waveguide radiating into a dielectric slab, *IEEE Transactions on Antennas and Propagation*, **15** (99), pp. 627-33.

Croswell, W.F., Taylor, W.C., Swift, C.T. *et al.* (1968) The admittance of rectangular waveguide-fed aperture under an inhomogeneous plasma: Theory and experiment, *IEEE Transactions on Antennas and Propagation*, **16** (7), pp. 475-87.

Decreton, M.C. and Gardiol, F.E. (1974) Simple nondestructive method for measurement of complex permittivity, *IEEE Transactions on Instrumentation and Measurements*, **23** (12), pp. 434-38.

Galejs, J. (1965) Admittance of a waveguide radiating into stratified plasma, *IEEE Transactions on Antennas and Propagation*, **13** (1), 64-70.

Galejs, J. (1969) Antennas in inhomogeneous media, Pergamon Press, Elmstord, New York.

Ganchev, S., Bakhtiari, S. and Zoughi, R. (1992) A novel numerical technique for dielectric measurement of lossy dielectrics, *IEEE Transactions on Instrumentation and Measurement*, **41** (6), pp. 361-65.

Ganchev, S., Qaddoumi, N., Bakhtiari, S. *et al.* (1995a) Calibration and measurement of dielectric properties of finite thickness composite sheets with open-ended coaxial sensors, *IEEE Transactions on Instrumentation and Measurement*, **44** (6), pp. 1023-29.

Ganchev, S., Qaddoumi, N., Ranu, E. *et al.* (1995b) Microwave detection optimization of disbond in layered dielectrics with varying thicknesses, *IEEE Transactions on Instrumentation and Measurement*, **44** (2), pp. 326-28.

Ganchev, S., Runser, R., Qaddoumi, N. *et al.* (1995c) Microwave nondestructive evaluation of thick sandwich composites, *Materials Evaluation*, **53** (4), pp. 389-95.

Gray, S. and Zoughi, R. (1997) Dielectric sheet thickness variation and disbond detection in multi-layered composites using an extremely sensitive microwave approach, *Materials Evaluation*, **55** (1), pp. 42-48.

Han, H.C. and Mansueto, E.S. (1995) Thin film inspection with millimeter-wave reflectometry, *Research in Nondestructive Evaluation*, **7** (2/3), pp. 98-100.

Han, H.C. and Mansueto, E.S. (1997) Thickness measurement for thin films and coatings using millimeter waves, *Research in Nondestructive Evaluation*, **9** (2), pp. 97-118.

Harrington, R.F. (1961) Time Harmonic electromagnetic fields, McGraw-Hill, New York, p. 381.

Jamieson, A.R. and Rozzi, T.E. (1977) Rigorous analysis of crosspolarization in flangemounted waveguide radiator, *Electronic Letters*, **13** (11), pp. 742-44.

Lewin, L. (1951) Advanced theory of waveguides, Illife, Ltd., London.

MacPhie, R.H. and Zaghloul, A.I. (1980) Radiation from a rectangular waveguide with infinite flange-Exact solution by the correlation matrix method, *IEEE Transactions on Antennas and Propagation*, **28** (7), pp. 497-503.

Nikita, K.S. and Uzungolu, N.K. (1989) Analysis of the power coupling from a waveguide hyperthermia applicator into a three-layered tissue model, *IEEE Transactions on Microwave Theory and Techniques*, **37** (11), pp. 1794-1801.

Qaddoumi, N., Handjojo, L., Bigelow, T. *et al.* (1999, in-press) Microwave corrosion detection using open-ended rectangular waveguide sensors, *Materials Evaluation.*

Qaddoumi, N., Shroyer, A. and Zoughi, R. (1997) Microwave Detection of Corrosion Under Paint and Composite Laminate Coatings. *Research in Nondestructive Evaluation*, **9** (4), pp. 201-212.

Qaddoumi, N., Zoughi, R. and Carriveau, C.W. (1996) Microwave detection and depth determination of disbonds in low-permittivity and low-loss thick sandwich composites, *Research in Nondestructive Evaluation*, **8** (1), pp. 51-63.

Teodoridis, V., Sphicopoulos, T. and Gardiol, F.E. (1985) The reflection from an open-ended rectangular waveguide terminated by layered dielectric medium, *IEEE Transactions on Microwave Theory and Techniques*, **33** (5).

Ulaby, F.T., Moore, R.K. and Fung, A.K. (1981) Microwave remote sensing, active and passive, Vol. I, Chapter 4, Artech House, Dedham, MA.

Villeneuve, A.T. (1965) Admittance of a waveguide radiating into plasma environment, *IEEE Transactions on Antennas and Propagation*, **13** (1), 115-21.

Zoughi, R. and Bakhtiari, S. (1990) Microwave nondestructive detection and evaluation of disbonding and delamination in layered dielectric slabs, *IEEE Transactions on Instrumentation and Measurement*, **39** (4), pp. 1059-63.

Zoughi, R. and Lujan, M. (1990) Nondestructive microwave thickness measurement of dielectric slabs, *Materials Evaluation*, **48** (10), pp. 1100-05.

Zoughi, R. (1995) Microwave and millimeter wave nondestructive testing: A succinct introduction, *Research in Nondestructive Evaluation*, **7** (2/3), pp. 71-74.

Zoughi, R. and Zonnefeld, B. (1991) Permittivity characteristics of Kevlar, carbon composites, fiberglass and rubber (33% Carbon) at X-band (8-12 GHz), *Proceedings of the Review of Progress in Quantitative NDE*, **10B**, pp. 1431-36, Plenum Press, New York.

CHAPTER 4

Surface crack evaluation

4.1 INTRODUCTION[1]

Metal fatigue or failure usually begins from the surface. Aircraft fuselage, nuclear power plant steam generator tubings and steel bridges are examples of environments in which this type of metal failure occurs. Hence, fatigue and stress crack detection on metallic structures is of utmost importance to the on-line and in-service inspections of metallic components. Currently, there are several prominent non-destructive testing (NDT) techniques for detecting surface cracks in metals; however, each method possesses certain limitations and disadvantages. In some environments the technique used may not be an optimum one, but the only one that can be applied. Acoustic emission testing, dye penetrant testing, eddy current testing, ultrasonic testing, radiographic testing, and magnetic particle testing are examples of these techniques (Bovig, 1989).

Since the late sixties there have been several researchers who have attempted using microwaves for surface crack detection on metals. Microwave techniques offer certain advantages, when detecting hairline stress or fatigue cracks, such as: the sensor may or may not be in contact with the surface under examination; they are applicable in high-temperature environments; crack may be filled with dielectric materials such as dirt, paint or rust; the surface of the metal may be covered with paint or a similar compound and the crack may still be detected. Finally, polarization properties of microwaves can provide information regarding relative crack orientation. Microwave techniques have also shown the potential of estimating crack width, depth and length (Zoughi *et al.*, 1995; Qaddoumi *et al.*, 1998b).

---

[1] Portions of this section are reprinted with permission from (Transactions on Instrumentation and Measurement, IM-43, no. 5, pp. 719-25, October 1994).

Some investigators used a mode conversion technique based on the idea that the crack converts a portion of the incident wave to an orthogonally polarized wave (Feinstein and Hruby, 1967; Feinstein and Hruby, 1968; Hruby and Feinstein, 1970). This non-contact technique utilized a microwave bridge for nulling out background signal, and a microwave rotary joint for producing incident waves of different polarizations. They were able to detect cracks with widths of 0.05 mm and different depths. The drawbacks of this technique are the introduction of the additional loss associated with the microwave bridge and the low-frequency mechanical modulation associated with the rotary joint. Bahr used a similar technique at 100 GHz (Bahr, 1981). He used mode conversion without polarization modulation. To separate the orthogonally polarized wave from the co-polarized backscattered wave an orthomode coupler was utilized. To increase the spatial resolution of the measurement apparatus, he used a focusing lens on a horn antenna to create a beamwidth equivalent to 3.5 mm at the focal point. The integrity of this approach was checked by examining 0.25 mm wide cracks on aluminum plates. He showed that at high enough frequencies, the depth of a crack may also be determined. The disadvantage of this method is that detection is directly dependent on the degree of decoupling between the orthogonally polarized signals created by the mode conversion in the crack. He also used circularly polarized signals and a dielectric waveguide to improve detection sensitivity such that fatigue cracks under loading were detected (Bahr, 1980). Other microwave approaches have included microstrip planar lines and ferromagnetic resonance probes for crack detection (Robinson and Gysel, 1972; Gysel and Feinstein, 1974; Husain and Ash, 1975, Bahr and Watjen, 1981; Ash and Husain, 1973; Auld, 1978; Auld, 1981).

4.2 OPEN-ENDED WAVEGUIDE APPROACH[2]

In mid-1992 several experiments, using an open-ended rectangular waveguide, were conducted to investigate the feasibility of using this probe to detect long surface cracks in metals (Yeh and Zoughi, 1994a). In this context, long refers to a crack whose length is greater than or equal to the broad dimension of a waveguide. Various long and thin slots of different widths and depths were sawcut in flat aluminum plates using jeweler saws. Throughout this chapter the terms thin slot or slot and crack are interchangeably used. Furthermore, unless otherwise mentioned all cracks

---

[2] Portions of this section are reprinted with permission from (Transactions on Instrumentation and Measurement, IM-43, no. 5, pp. 719-25, October 1994). © 1994 IEEE.

used in this chapter have been cut out of aluminum plates. Preliminary experiments were conducted by moving (using a computer-controlled stepping motor) a cracked metal surface over the aperture of an open-ended rectangular waveguide while monitoring the standing wave characteristics inside the waveguide. Subsequently, it was observed that when the crack axis (length) is parallel to the broad dimension of the waveguide (i.e. when the length is orthogonal to the electric field of the dominant TE_{10} mode) the standing wave experiences a pronounced shift in location when the crack is exposed to the aperture of the waveguide compared to when the crack is outside the aperture (i.e. a short circuit condition). This shift indicates changes in the reflection coefficient properties of the metal surface perturbed by the crack. It was also observed that this shift is highly dependent on the relative location of the crack within the waveguide aperture (i.e. whether the crack is at the edge or at the center of the aperture). Figures 4.1 and 4.2 show the side- and plan- views of a crack with width (or opening) W, depth D and length L and a waveguide aperture with dimensions a and b, when the crack length is parallel to the broad dimension of the waveguide. δ is a dimension indicating the location of the crack relative to an arbitrary location on the small dimension of the waveguide aperture, b. This parameter will be referred to as the scanning distance throughout this chapter. It was also observed that when the crack was not parallel to the broad dimension of the waveguide, the level of change in the standing wave decreased, and when the crack became parallel to the smaller dimension of the waveguide (parallel to the dominant TE_{10} mode electric field vector) there was no measurable perturbation in the characteristics of the standing wave. This is due to the fact that in this case the surface currents on the metal surface, induced by the probing waveguide, are parallel to the crack length which does not perturb the surface currents.

Figure 4.3 shows a simple measurement apparatus that was used for these experiments. An oscillator feeds a (slotted) waveguide terminated by a metal plate in which there exists a crack. The diode detector is placed a distance ℓ away from the waveguide aperture, where the metal plate is scanned over the waveguide aperture and a voltage proportional to the generated standing wave pattern inside the waveguide is recorded. As will be seen later, different detector locations, ℓ, will change the difference between the measured signals for the short circuit (i.e. crack outside of the waveguide aperture) case and when the crack is in the middle of the aperture. If ℓ is chosen such that the detector is located between a maximum and a minimum on the standing wave pattern, this difference is substantial (Yeh, 1994; Yeh and Zoughi, 1994a, Fig. 6).

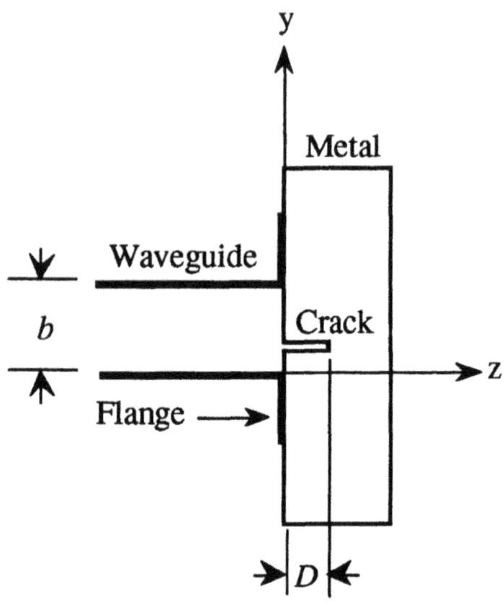

Fig. 4.1 Side-view of a surface crack and an open-ended rectangular waveguide aperture (Yeh and Zoughi, 1994a).

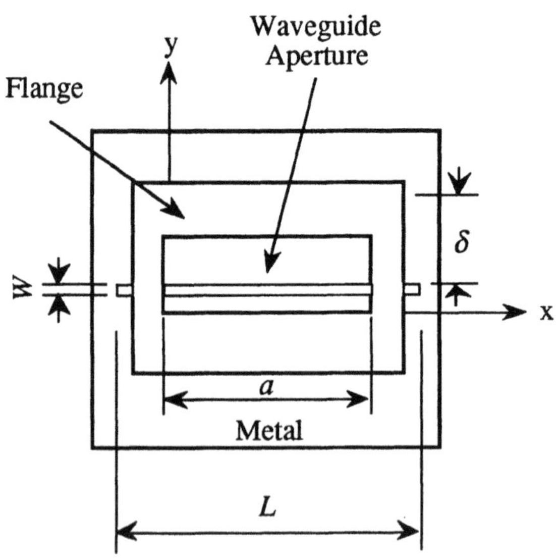

Fig. 4.2 Plan-view of a surface crack and an open-ended rectangular waveguide aperture (Yeh and Zoughi, 1994a).

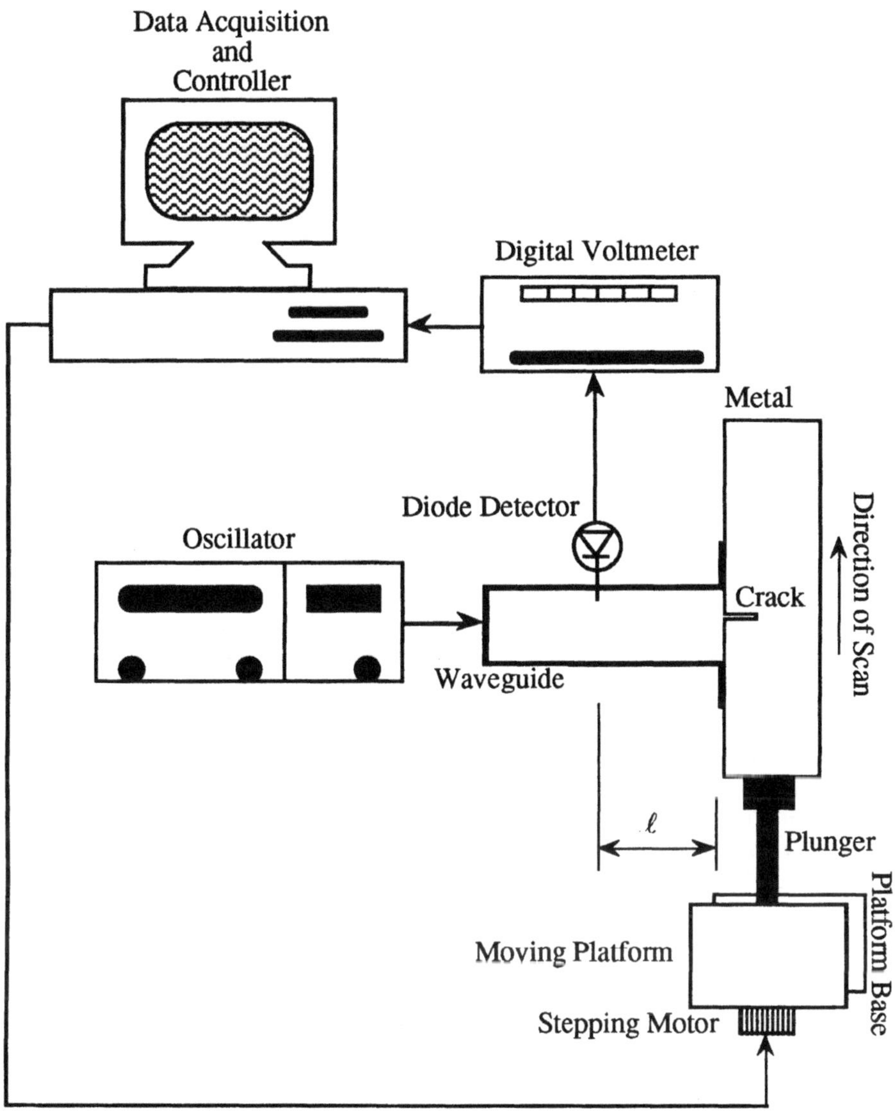

Fig. 4.3 Laboratory apparatus for surface crack evaluation (Yeh and Zoughi, 1994a).

At a frequency of 24 GHz, a long crack with a length of $L > 10.67$ mm, a width of 0.84 mm and a depth of 1.03 mm was scanned over the aperture of a K-band waveguide for which $a = 10.67$ mm and $b = 4.32$ mm. The diode output voltage measured at $\ell = 9.45$ cm is shown in Fig. 4.4. The results indicate that while the crack is outside the waveguide aperture the diode registers very little voltage variation due to the fact that the waveguide is terminated by a fairly good short-circuited load. The noise-like feature associated with the signal is due to the quantization resolution of the A/D converter and the internal noise of the voltmeter. As the crack begins to appear within the waveguide aperture the voltage experiences a rapid magnitude change which is an indication of rapid phase change in the reflection coefficient at the aperture. The same phenomenon occurs when the crack leaves the waveguide aperture. The voltage value does not change very much while the crack is inside the aperture; however, its value is still different than that of the short circuit case. The diode output voltage as a function of δ is clearly an indication of the presence of the crack (detection) and is hereon referred to as the **crack characteristics signal**, since the absence of the crack results in a fairly constant voltage. Furthermore, as will be discussed later, for a given frequency of operation and waveguide dimensions the distance between the two sharp transitions is primarily dependent upon the crack width plus the narrow dimension of the waveguide. The voltage level difference between when the crack is outside the waveguide aperture (the short circuit case) and when the crack is inside the waveguide aperture is primarily a function of the crack depth and to a lesser extent a function of its width. Therefore, the crack characteristic signal is also a means for evaluating the dimensions of a crack (evaluation). This issue will be discussed in detail later.

Important to the theoretical calculations of the crack characteristic signal, it was experimentally determined that a thin slot or crack with a length equal to the broad dimension of the waveguide produces the same crack characteristic signal as one with a length greater than the broad dimension of the waveguide such as the one shown in Fig. 4.4 (Yeh and Zoughi, 1994a, Fig. 3). The important ramification of this observation is that a thin slot may be considered as a short-circuited waveguide with the same broad dimension as the probing waveguide and with a narrow dimension equal to the width or opening of the crack. Therefore, the dominant mode electric field properties in the probing waveguide and the crack are identical, since the cutoff frequency of this mode is determined by the broad dimension of the waveguide in which it propagates.

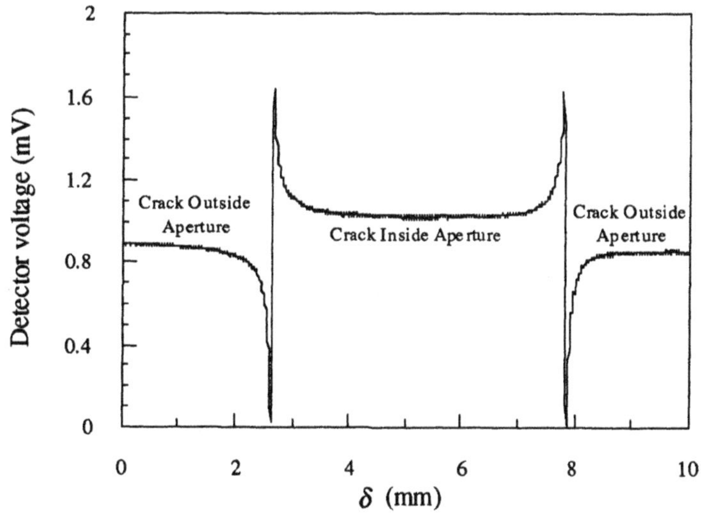

Fig. 4.4 Experimental crack characteristic signal for a long crack with width W = 0.84 mm and depth D = 1.03 mm at a frequency of 24 GHz (Yeh and Zoughi, 1994a).

4.3 THEORETICAL ANALYSIS FOR EXPOSED CRACKS[3]

Two different approaches have been proposed to model the interaction of a crack and a waveguide aperture (Yeh, 1994; Huber 1996). In both approaches long surface cracks are assumed to be narrow slots cut out of a metal plate, and since it has been shown that long cracks and cracks with a length equal to the broad dimension of the probing waveguide produce very similar crack characteristic signals, in both approaches it was assumed that a long surface crack has a length equal to the broad dimension of the probing waveguide (Yeh and Zoughi, 1994a, Fig. 3). In this way the problem reduces to the larger probing waveguide interacting with a smaller waveguide (e.g. the crack). Since the crack opening or width is assumed to be small, given the incident dominant TE_{10} mode at the waveguide aperture, in the first approach it was assumed that the presence of the crack within the aperture only produces higher-order reflected TM modes (Yeh and Zoughi, 1994a). This assumption is quite reasonable for tight cracks. However, its validity diminishes as the crack width increases and when the length of the crack is less than the broad dimension of the probing waveguide aperture

[3] Portions of this section are reprinted with permission from (Transactions on Microwave Theory and Techniques, MTT-45, no. 4, pp. 477-84, April 1997). © 1997 IEEE.

(Yeh and Zoughi, 1994b). Nevertheless, for long cracks this assumption reduces the analytical complexity of the modeling effort considerably. In addition, this model relies on a mode matching approach to analyze the electromagnetic properties of a system formed by the probing waveguide, and the crack as a function of crack location within the waveguide aperture. The inherent drawback of this approach is evident when the crack is at the edge of the waveguide aperture in which case many modes are needed to replicate the sharp transitions. Hence, this model which in effect is a brute force approach of setting the boundary conditions and solving for the unknown coefficients becomes increasingly computer resource intensive. This problem is more severe for finite cracks in which both the higher-order reflected TM and TE modes need to be considered. In addition, the solution is crack location dependent. Finally, this model is not general and may be only applied to finite cracks after substantial modification. Nevertheless, this first approach has significantly aided in the understanding of the interaction of a surface crack modeled as a small waveguide fed by a larger probing waveguide aperture (Yeh, 1994; Yeh and Zoughi, 1994a).

The second approach evaluates the change in the reflection coefficient properties of a generalized system encompassing empty, filled and finite cracks, located at an arbitrary position inside the probing waveguide aperture (Huber, 1996; Huber *et al.*, 1997a). A magnetic current density, $\boldsymbol{M}$, is introduced over the common aperture formed by the waveguide and the crack. Subsequently, the junction formed by the waveguide and the cracked metallic surface is separated into two systems. A numerical solution employing the method of moments is obtained, and the reflection coefficient at the waveguide aperture is expressed in terms of a generalized scattering matrix. The convergence behavior is studied to determine an optimized set of basis functions and the optimal number of higher-order modes required for a fast and accurate solution. Since this modeling approach is more versatile and provides for a more convenient convergence analysis for examining the appropriateness of the incorporated number of higher-order modes, it is the one presented in this chapter. The reader is encouraged to examine the first model as a reference for better understanding of the current modeling approach and its advantages.

In order to obtain a general representation of a system formed by a waveguide aperture and a metallic surface with a crack, arbitrary incident electric and magnetic fields in the waveguide are assumed. The incident and reflected fields in the waveguide and the crack are expressed in terms of their discrete orthonormal eigenfunctions (for the dominant mode and the higher-order modes) with unknown complex coefficients. These coefficients represent the amplitudes and the phases of the respective eigenfunctions. A magnetic current density, $\boldsymbol{M}$, is introduced over the common aperture of the system formed by the waveguide and the crack, as shown in Figs. 4.5 and 4.6. This system can then be separated into two parts. It must be noted that in this approach an analysis of the electromagnetic properties as a function of the relative crack location within the waveguide aperture is needed. This means, that it is necessary to evaluate the change in the reflection coefficient as the crack is being scanned (i.e. the crack location is continuously varying within the waveguide aperture). This includes the crack being partially outside in either x- or y-directions. Applying the method of moments provides for a numerical solution for the complex field coefficients (Harrington, 1961). The accuracy of approximating the electric and magnetic field distributions anywhere in the waveguide or in the crack subsequently depends upon the number of higher-order modes used, and it depends on the appropriate choice of the basis functions for the method of moments. The convergence behavior is then used for analyzing all of these criteria. Finally, a generalized scattering matrix is formulated by writing the system of equations in a matrix form and solving for the reflection coefficient at the aperture of the waveguide (Harrington and Mautz, 1976).

4.3.1 Formulation of the generalized scattering matrix

The fields in the waveguide and the crack are represented by their orthonormal mode vectors which form complete sets for describing the respective electromagnetic fields (Huber, 1996; Borgnis and Papas, 1958; Marcuvitz, 1951). By normalization the power is divided between modes according to the square of their amplitudes (Collin, 1992). The *i*th orthonormal mode vectors for the waveguide side (subscript *w*) are given by $\boldsymbol{e}_{wi}^{\mathrm{TE}}$, $\boldsymbol{h}_{wi}^{\mathrm{TE}}$, $\boldsymbol{e}_{wi}^{\mathrm{TM}}$ and $\boldsymbol{h}_{wi}^{\mathrm{TM}}$. For the crack side (subscript *c*) the orthonormal mode vectors $\boldsymbol{e}_{ci}^{\mathrm{TE}}$, $\boldsymbol{h}_{ci}^{\mathrm{TE}}$, $\boldsymbol{e}_{ci}^{\mathrm{TM}}$ and $\boldsymbol{h}_{ci}^{\mathrm{TM}}$ are given in a similar fashion by taking the appropriate physical dimensions into account (replacing *a* with *L* and *b* with *W*, respectively). Bold characters indicate vector or matrix quantities. The orthonormal mode vectors satisfy the orthogonality relationship in the waveguide and in the crack, respectively. The fields in the waveguide and the crack are well defined by the solutions of Maxwell's equations that satisfy all of the boundary conditions except at the junction. Forcing the boundary conditions for the transverse fields at the aperture renders an effective solution for all of the unknown field coefficients.

For a general representation, the system formed by the waveguide and the crack is divided into two parts using the equivalence principle. The equivalence principle states that the fields in the waveguide are identical to the excitation fields plus the fields produced by an equivalent magnetic current density, $\boldsymbol{M}$, when the aperture *S* is replaced by a perfect conductor (Harrington, 1961).

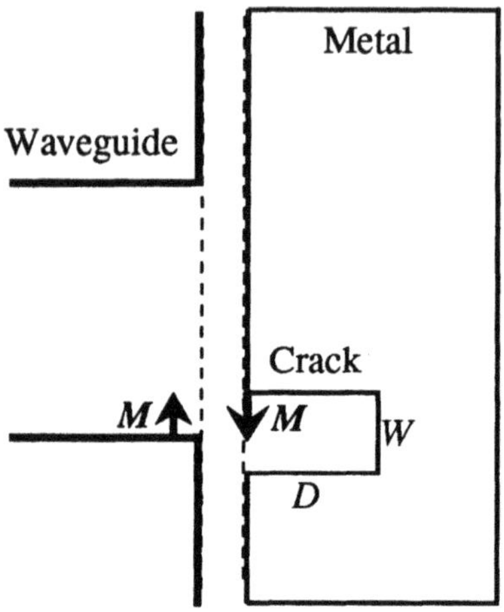

Fig. 4.5 Side-view of the relative geometry of a surface crack and a waveguide aperture (Huber *et al.*, 1997a).

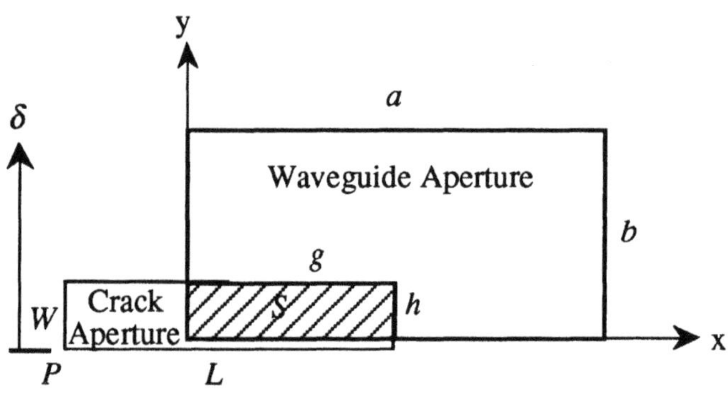

Fig. 4.6 Plan-view of the relative geometry of a surface crack, a waveguide aperture and the coordinate system where z-axis is out of the page (Huber *et al.*, 1997a).

In the crack, the total field is composed of two components resulting from the reflection by the short-circuited end of the crack plus the field produced by the equivalent magnetic current density, -***M***, over the aperture *S*, as shown in Fig. 4.6.

Hence, the total transverse electric and magnetic fields in the waveguide are now given by:

$$\boldsymbol{E}_{wt} = \sum_i C_i e^{-\gamma_{wi} z} \boldsymbol{e}_{wi} - \sum_i C_i e^{\gamma_{wi} z} \boldsymbol{e}_{wi} + \sum_i D_i e^{\gamma_{wi} z} \boldsymbol{e}_{wi} \quad (4.1a)$$

$$\boldsymbol{H}_{wt} = \sum_i C_i Y_{wi} e^{-\gamma_{wi} z} \boldsymbol{u}_z \times \boldsymbol{e}_{wi} + \sum_i C_i Y_{wi} e^{\gamma_{wi} z} \boldsymbol{u}_z \times \boldsymbol{e}_{wi}$$

$$- \sum_i D_i Y_{wi} e^{\gamma_{wi} z} \boldsymbol{u}_z \times \boldsymbol{e}_{wi} \quad (4.1b)$$

Here, C_i and D_i are the respective coefficients of the incident modes and the modes produced by $\boldsymbol{M}$. In the crack the total transverse fields are then given by:

$$\boldsymbol{E}_{ct} = \sum_i B_i e^{-\gamma_{ci} z} \boldsymbol{e}_{ci} - \sum_i B_i e^{\gamma_{ci} z} \boldsymbol{e}_{ci} + \sum_i G_i e^{-\gamma_{ci} z} \boldsymbol{e}_{ci} \quad (4.2a)$$

$$\boldsymbol{H}_{ct} = \sum_i B_i Y_{ci} e^{-\gamma_{ci} z} \boldsymbol{u}_z \times \boldsymbol{e}_{ci} + \sum_i B_i Y_{ci} e^{\gamma_{ci} z} \boldsymbol{u}_z \times \boldsymbol{e}_{ci}$$

$$+ \sum_i G_i Y_{ci} e^{-\gamma_{ci} z} \boldsymbol{u}_z \times \boldsymbol{e}_{ci} \quad (4.2b)$$

with B_i and G_i being the respective coefficients of the reflected modes and the modes produced by -$\boldsymbol{M}$. The last terms in equations (4.1) and (4.2) corresponds to the fields generated by the equivalent magnetic current density, $\boldsymbol{M}$ (i.e. at z = 0 the first two terms cancel each other). γ_{qi} is the mode propagation constant and Y_{qi} is the modal characteristic admittance in the waveguide and in the crack, $q \in \{w, c\}$ (Huber, 1996).

4.3.2 Application of the boundary conditions

For a general case as shown in Figs. 4.5 and 4.6, the transverse electric fields must vanish at the short-circuited end of the crack, z = D; thus, it follows that:

$$G_i = B_i \left[e^{2\gamma_{ci} D} - 1 \right] \quad (4.3)$$

Next, the continuity of the transverse electric field $\boldsymbol{E}_t$ across the common aperture, S, has to be satisfied. The placement of an equivalent magnetic current density, +$\boldsymbol{M}$, across S in the waveguide and -$\boldsymbol{M}$ across S in the crack ensures the continuity of $\boldsymbol{E}_t$ across this common aperture. The equivalent magnetic current density, $\boldsymbol{M}$, can hence be evaluated from equations (4.1) and (4.2) as:

$$M = u_z \times E_{wt}\Big|_{z=0} = \sum_i D_i u_z \times e_{wi} \tag{4.4a}$$

and,

$$M = u_z \times E_{ct}\Big|_{z=0} = \sum_i B_i\left[e^{2\gamma_{ci}D} - 1\right] u_z \times e_{ci} \tag{4.4b}$$

Likewise, the continuity of H_t across the aperture S requires that:

$$2\sum_i C_i Y_{wi} u_z \times e_{wi} = \sum_i D_i Y_{wi} u_z \times e_{wi} + \sum_i B_i Y_{ci}\left[e^{2\gamma_{ci}D} + 1\right] u_z \times e_{ci} \tag{4.5}$$

In order to obtain a numerical solution for equation (4.5), the method of moments is employed (Harrington, 1961). u_x, u_y and u_z are the coordinate system unit vectors.

4.3.3 Application of the method of moments

M is first expanded as a complete set of basis functions, M_j, which consists of N real valued expansion functions, and V_j are complex variables which are unknown and must be determined. These basis functions, which describe the behavior of the magnetic field right at the aperture, must be chosen appropriately for obtaining a close-to-the-exact solution and to achieve fast convergence. If the number of modes in the waveguide is limited to N_w and the number of modes in the crack to N_c, and N expansion functions are used, then:

$$\sum_{j=1}^{N} V_j M_j = \sum_{i=1}^{N_w} D_i u_z \times e_{wi} \tag{4.6a}$$

$$\sum_{j=1}^{N} V_j M_j = \sum_{i=1}^{N_c} B_i\left[e^{2\gamma_{ci}D} - 1\right] u_z \times e_{ci} \tag{4.6b}$$

Now, by using orthogonality of mode vectors, and by scalarly multiplying equations (4.6a) and (4.6b) by $(u_z \times e_{wk})$ and by $(u_z \times e_{ck})$ respectively, and then integrating over the corresponding waveguide and crack aperture surfaces, S_w and S_c, the following equations are obtained:

$$D_i = \sum_{j=1}^{N} V_j H_{wij} \qquad (i = 1, 2, \cdots, N_w) \tag{4.7a}$$

$$B_i = \left(\sum_{j=1}^{N} V_j H_{cij} \right) \left[e^{2\gamma_{ci} D} - 1 \right]^{-1} \qquad (i = 1, 2, \cdots, N_c) \tag{4.7b}$$

where,

$$H_{wij} = \iint_{S_w} \boldsymbol{M}_j \cdot \boldsymbol{u}_z \times \boldsymbol{e}_{wi} \, dS_w \tag{4.8a}$$

$$H_{cij} = \iint_{S_c} \boldsymbol{M}_j \cdot \boldsymbol{u}_z \times \boldsymbol{e}_{ci} \, dS_c \tag{4.8b}$$

Using Galerkin's method, and forming the scalar product of equation (4.5) with each of the testing functions $\boldsymbol{M}_k$ gives:

$$2 \sum_{i=1}^{N_w} C_i Y_{wi} P_{wik} = \sum_{i=1}^{N_w} D_i Y_{wi} P_{wik} + \sum_{i=1}^{N_c} B_i Y_{ci} \left[e^{2\gamma_{ci} D} + 1 \right] P_{cik} \tag{4.9}$$

where,

$$P_{wik} = \iint_{S_w} \boldsymbol{M}_k \cdot \boldsymbol{u}_z \times \boldsymbol{e}_{wi} \, dS_w \tag{4.10a}$$

$$P_{cik} = \iint_{S_c} \boldsymbol{M}_k \cdot \boldsymbol{u}_z \times \boldsymbol{e}_{ci} \, dS_c \tag{4.10b}$$

After substitution for D_i and B_i from equation (4.7), the following equation is obtained:

$$2 \sum_{i=1}^{N_w} C_i Y_{wi} P_{wik} = \sum_{i=1}^{N_w} \left(\sum_{j=1}^{N} V_j H_{wij} \right) Y_{wi} P_{wik} + \sum_{i=1}^{N_c} \left(\sum_{j=1}^{N} V_j H_{cij} \right) \frac{\left[e^{2\gamma_{ci} D} + 1 \right]}{\left[e^{2\gamma_{ci} D} - 1 \right]} Y_{ci} P_{cik} \tag{4.11}$$

Equation (4.11) is now written in a matrix form as:

$$\vec{I} = \left[\overline{Y}_w + \overline{Y}_c\right]\vec{V} \tag{4.12}$$

where,

$$\vec{I} = 2P_w^T Y_w \vec{C} \tag{4.13a}$$

$$\overline{Y}_w = P_w^T Y_w H_w \tag{4.13b}$$

$$\overline{Y}_c = P_c^T Y_c E_1 E_2 H_c \tag{4.13c}$$

Y_w, Y_c, E_1 and E_2 are diagonal matrices whose elements are readily evaluated (Huber, 1996). Note that $P_w^T = H_w^T$ and $P_c^T = H_c^T$ when using the Galerkin's method.

4.3.4 Generalized scattering matrix

Now it is possible to evaluate the generalized scattering matrix, S, of the waveguide-crack junction. From equation (4.1) the reflected wave is given by:

$$\vec{A} = \vec{D} - \vec{C} \tag{4.14}$$

Substituting for $\vec{D}$ and evaluating $\vec{V}$ from equation (4.12) gives:

$$\vec{D} = H_w \vec{V} = H_w 2\left[\overline{Y}_w + \overline{Y}_c\right]^{-1} P_w^T Y_w \vec{C} \tag{4.15}$$

Then, equation (4.14) is rewritten as:

$$\vec{A} = \left[\, 2H_w\left[\overline{Y}_w + \overline{Y}_c\right]^{-1} P_w^T Y_w - U \right]\vec{C} \tag{4.16}$$

where U is the unity matrix. Subsequently, S_{11} is expressed as (Huber, 1996):

$$S_{11} = \vec{A}\vec{C}^{-1} = 2H_w\left[\overline{Y}_w + \overline{Y}_c\right]^{-1} P_w^T Y_w - U \tag{4.17}$$

Note that S_{ij} is the amplitude of the ith mode due to the jth incident mode with unit amplitude. Here, we are not concerned with S_{21}, S_{12} and S_{22} which are the other scattering coefficients in the waveguide and the crack.

However, if need be, they can be readily calculated as well.

In order to calculate the crack characteristic signal, it is necessary to evaluate the sub-matrix S_{11} (reflection coefficient of the dominant mode and each of the higher-order modes) of the generalized scattering matrix, **S**, for different positions of the crack relative to the waveguide aperture. Once the reflection coefficient of the incident dominant TE_{10} mode is known, the shift of the standing wave in the waveguide can be evaluated. Subsequently, the diode output voltage (probing the standing wave) for the dominant mode detection technique can be computed to obtain the crack characteristic signal (Yeh and Zoughi, 1994a).

4.3.5 Convergence

The above moment solution approach and the subsequent generalized scattering parameters depend on the choice of an initially unknown equivalent magnetic current density, **M**, over a conducting surface. **M** is described by basis functions that form a complete set. Since **M** describes the physical behavior of the magnetic field over the common aperture *S*, it is best to choose the basis functions to have similar properties to the orthonormal mode vectors of the transverse magnetic fields in the waveguide and the crack. In this way, a relatively fast convergence may be obtained. Hence, **M** is chosen as following:

$$\boldsymbol{M}_j = \boldsymbol{u}_x \sin\left(\frac{q'\pi x}{g}\right)\cos\left(\frac{p'\pi y}{h}\right) + \boldsymbol{u}_y \cos\left(\frac{q\pi x}{g}\right)\sin\left(\frac{p\pi y}{h}\right) \qquad (4.18)$$

where *g* and *h* are the dimensions of the common aperture opening *S* (Fig. 4.6). It must be noted that *q'* and *p* cannot be equal to zero. This is obvious from the fact that TE_{0m} and TE_{n0} modes exist, but not TM_{0m} or TM_{n0} modes.

The rate of convergence is critically dependent on the number of basis functions used to describe **M**. Figure 4.7 shows that faster convergence is achieved when the maximal indices in x- and y-direction are chosen to be the same ($q'_{max} = q_{max}$ and $p'_{max} = p_{max}$). The normalized signal (i.e. the diode output voltage representing one point of the crack characteristic signal) is calculated and plotted for an increasing number of modes in the waveguide for a crack at the relative coordinates of (x = 0 mm, y = -0.4 mm) at 24 GHz (K-band waveguide dimensions are *a* = 10.67 mm and *b* = 4.32 mm). For simplicity, a crack with its length equal to the broad dimension of the waveguide is considered (i.e. a long crack). Thus, only TE_{1m} and TM_{1m} modes have to be considered, and $q'_{max} = q_{max} = 1$, as no variation in the x-direction is encountered (Yeh and Zoughi, 1994a). The

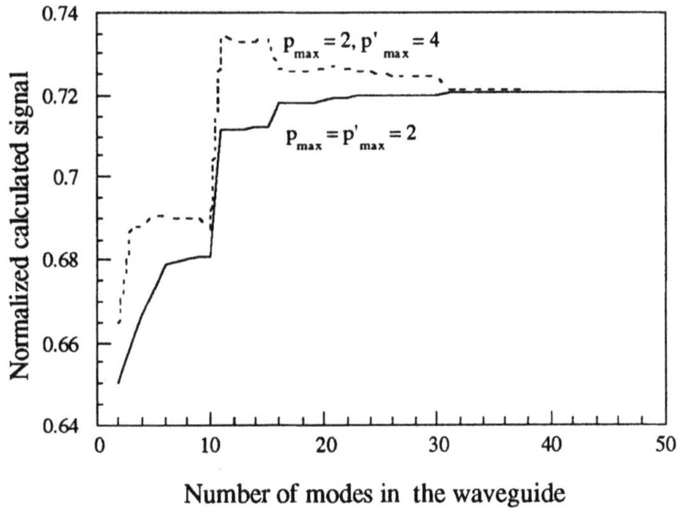

Fig. 4.7 Convergence curves for a long crack with width W = 0.84 mm and depth D = 1.53 mm at 24 GHz, at the relative coordinates (x = 0 mm, y = -0.4 mm), with increasing numbers of basis functions in x- and y-directions (Huber *et al.* , 1997a).

number of modes in the crack is chosen proportional to the number of modes in the waveguide by the ratio of W over b. In the case of $p'_{max} = p_{max} = 2$, the final value for the diode output voltage is achieved with 22 modes in the waveguide compared to 36 modes for $p'_{max} = 4$, $p_{max} = 2$, when allowing an error of 0.1%. Moreover, the overshooting effect as seen in Fig. 4.7 for $p'_{max} \neq p'_{max}$ is avoided. Allowing a larger error with respect to the final value would further significantly reduce the number of modes needed in the calculations.

In the case that the number of basis functions used to describe the magnetic current density, ***M***, is not sufficient, the problem of relative convergence is encountered (Uher, Bornemann and Rosenberg, 1993). Figure 4.8 shows the convergence curves for increasing numbers of basis functions. It is seen that once a sufficient number of basis functions is used, the signal amplitude of the crack characteristic signal does not change significantly (0.22% error for $p'_{max} = p_{max} = 1$, and 0.054% error for $p'_{max} = p_{max} = 2$). Increasing the number of basis functions further reduces the speed of convergence. Thus, for an error of less than 0.1%, 22 modes are needed in the waveguide for $p'_{max} = p_{max} = 2$ vs. 56 modes for $p'_{max} = p_{max} = 6$. Clearly, a minimal number of basis functions is desired for reducing the computation time while achieving accurate results.

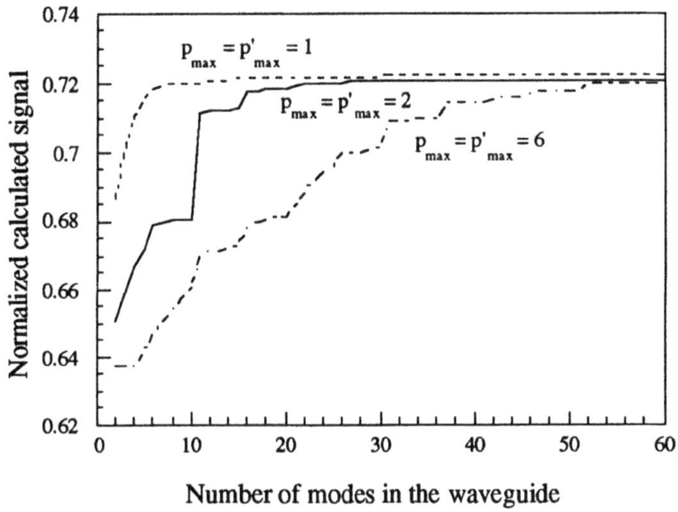

Fig. 4.8 Relative convergence problem for a long crack with width W = 0.84 mm and depth D = 1.53 mm at 24 GHz, at the relative coordinates (x = 0 mm, y = -0.4 mm), with increasing numbers of basis functions in x- and y-directions (Huber *et al.* , 1997a).

The optimal number of basis functions depends on the physical dimensions of the waveguide (a and b) and the crack (i.e. length L and width W), as well as the size of the common aperture, S (with dimensions g and h, which change as a function of the relative crack location within the waveguide aperture). Through extensive numerical evaluations the following criteria have been established for choosing an appropriate number of basis functions (Huber, 1996)

$$q'_{max} = q_{max} = \mathrm{ceil}\left(\frac{L}{g}\right) + 1 \leq 5 \quad (4.19a)$$

$$p'_{max} = p_{max} = \mathrm{ceil}\left(\frac{W}{h}\right) + 1 \leq 5 \quad (4.19b)$$

The operator (ceil) means rounding off to the larger integer. For achieving fast convergence with the method of moments it is best to approximate the actual physical behavior of the magnetic field over the aperture. In order to accommodate for the magnetic field at the edges of the aperture a different set of basis functions may be applied. For this case $\boldsymbol{M}$ is defined by:

$$M_j = u_x \frac{\sin\left(\frac{q'\pi x}{g}\right)\cos\left(\frac{p'\pi y}{h}\right)}{\sqrt{1-\beta\left(\frac{y-\frac{h}{2}}{\frac{h}{2}}\right)}}$$

$$+u_y \cos\left(\frac{q\pi x}{g}\right)\sin\left(\frac{p\pi y}{h}\right) \tag{4.20}$$

where β, a number close to one, is introduced to avoid infinity. The number of modes needed in the waveguide and the crack are not significantly reduced compared to that used in equation (4.18). On the contrary, the necessary numerical integrations for evaluating the matrix elements of H_{wij} and H_{cij} result in an overall slower computation time, and is therefore not considered.

4.3.6 Choice of higher-order modes

Once an optimal set of the basis functions is chosen, it becomes necessary to obtain a criterion for choosing an appropriate number of modes in the waveguide and in the crack. This number again depends on the physical dimensions of the waveguide, the crack and the size of the common aperture S (g and h). Since the evaluation of the crack characteristic signal is dependent on the position of the crack relative to the waveguide aperture, the size of the common aperture changes. When the crack is just about to enter the waveguide aperture, the common aperture, S, is very narrow (i.e. h being very small), and a larger number of modes is needed. A thorough study of the convergence behavior for the crack at different relative positions with respect to the waveguide aperture has indicated that the maximal indices n and m for the waveguide side are

$$n_{wmax} = \text{ceil}\left(\frac{q_{max}a}{g}\right) \tag{4.21a}$$

$$m_{wmax} = \text{ceil}\left(\frac{p_{max}b}{h}\right) \tag{4.21b}$$

The TE_{nm} and TM_{nm} modes are degenerate modes (modes with the same cutoff frequency) and hence must be considered as pairs (Collin, 1992).

Only in special situations it is possible to significantly reduce the number of modes used. For instance, for a long crack only TE_{1m} and TM_{1m} modes have to be considered as no variation in the x-direction is encountered (Yeh and Zoughi, 1994a). All the coefficients of the other higher-order modes would evaluate to zero in this special case.

For the number of modes in the crack a similar criterion is considered, and hence the maximal indices n_{cmax} and m_{cmax} of the desired number of TE_{nm} and TM_{nm} modes in the crack are given by:

$$n_{cmax} = \text{ceil}\left(\frac{q_{max}L}{g}\right) \tag{4.22a}$$

$$m_{cmax} = \text{ceil}\left(\frac{p_{max}W}{h}\right) \tag{4.22b}$$

It must be noted that when the crack is totally within the waveguide aperture ($L = g$ and $W = h$), the number of basis functions are identical to the number of modes in the crack. This is sufficient when M is chosen as in equation (4.18), because the basis functions over the common aperture, S, directly correspond to the orthonormal mode vectors of the magnetic field in the crack. Thus, the fields are described in a similar fashion, and the coefficients of all other higher-order modes in the crack would evaluate to zero (i.e. these modes are orthogonal to the basis functions).

4.3.7 Results

The tangential electric field E_y at the aperture of the waveguide (i.e. at z = 0) is evaluated numerically. Figure 4.9 shows the result for a finite crack/slot with a length of 6 mm, a width of 0.84 mm and a depth of 1.53 mm located at the relative coordinates (x = 2 mm, y = 1 mm). As expected, the tangential electric field E_y vanishes over the conducting surfaces, whereas it has a finite value over the common aperture, S. In order to further reduce the ripples of the electric field visible over the conducting surfaces more modes could be used.

To show the validity of this theoretical approach, the calculated crack characteristic signal is compared with the measured crack characteristic signal (Fig. 4.10) for a long milled crack/slot with a width of 0.55 mm and a depth of 2.5 mm at 24 GHz. The measured crack characteristic signal is normalized with respect to its short circuit value (crack outside the waveguide aperture). The agreement between the calculated and measured crack characteristic signal is very good.

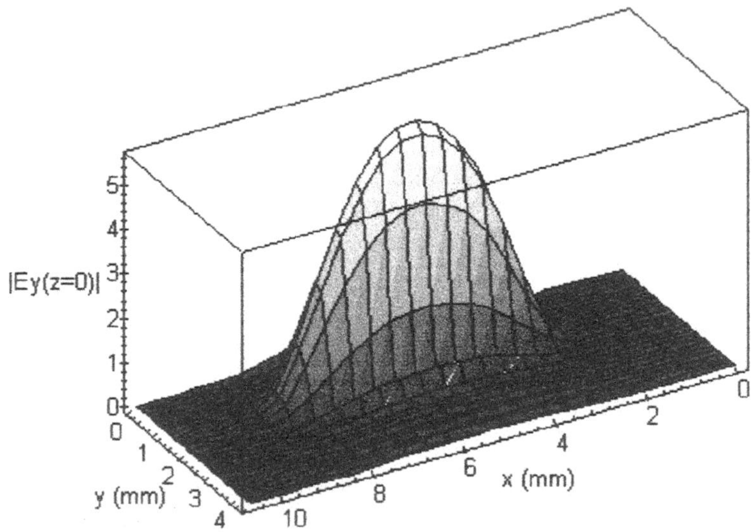

Fig. 4.9 Tangential electric field, E_y, at the waveguide aperture (z = 0) for a finite crack with a length of 6 mm, width W = 0.84 mm and depth D = 1.53 mm at 24 GHz (Huber *et al.*, 1997a).

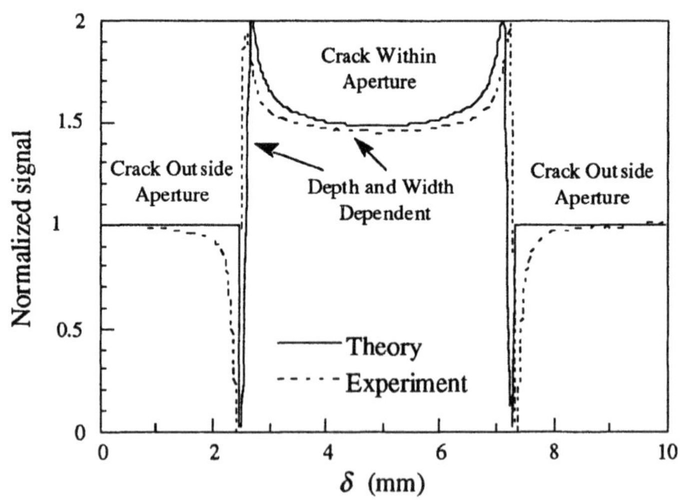

Fig. 4.10 Normalized crack characteristic signals for a long crack with width W = 0.55 mm and depth D = 2.5 mm at 24 GHz, calculated and measured (Huber *et al.*, 1997a).

The slight deviations are due to a limited number of modes used in the calculation, imperfection in machining a crack on a metal surface as specified in the calculations as well as the detector diode characteristics.

The properties of the crack characteristic signal are a function of the crack width, depth and length (Zoughi *et al.*, 1995, Qaddoumi *et al.*, 1998a; Qaddoumi *et al.*, 1998b). When the crack is filled with a dielectric material, its crack characteristic signal also changes compared to the case when it is empty. Figures 4.11 and 4.12 show the normalized calculated and measured crack characteristic signals, recorded at a frequency of 24 GHz, for empty and filled with rust powder (with measured dielectric property of $\varepsilon_r = 6.25 - j0.05$) cracks/slots, with a width of 0.85 mm and a depth of 1.25 mm, respectively. The reduction in the width of the crack characteristic signal (i.e. distance between the two sharp transitions) and the change in the middle level of the crack characteristic signal are evident, as experimentally investigated further in (Zoughi *et al.*, 1995). This method may also be used in tandem with other non-destructive inspection techniques (such as the dye penetrant method) for optimizing shallow crack depth determination.

4.4 HIGHER-ORDER MODE APPROACH[4]

Figures 4.1 and 4.2 show the side and plan view of the relative geometry of a crack and an open-ended waveguide aperture, respectively. In the absence of the crack only the dominant incident and reflected components of the electric field exists; namely, $E_y(z)$. However, the presence of a long crack generates an infinite number of reflected higher-order TM (Transverse Magnetic) modes in addition to the dominant mode (Yeh, 1994). However, for finite cracks this will no longer be true and the contribution of all higher-order modes must be taken into account (Yeh, 1994; Yeh and Zoughi, 1994b, Huber, 1996). Assuming a dominant TE_{10} mode incident on the metal surface, the reflected TE_{10} and the higher-order TM modes can be determined by forcing the boundary conditions at the metal surface and the boundaries inside the crack as the waveguide aperture scans over it. A brief explanation of the theoretical derivations is given here. The details of these derivations are given elsewhere (Yeh, 1994; Yeh and Zoughi, 1994a; Yeh, Ranu and Zoughi, 1994).

The dominant TE_{10} mode in the waveguide is the incident wave whose electric and magnetic fields (those components sufficient for deriving the expressions for the higher-order TM modes) are given by (Pozar, 1990):

---

[4] Portions of this section are reprinted with permission from *Materials Evaluation*, vol. 52, no. 6, pp. 676-81, Copyright 1994 © The American Society for Nondestructive Testing, Inc.

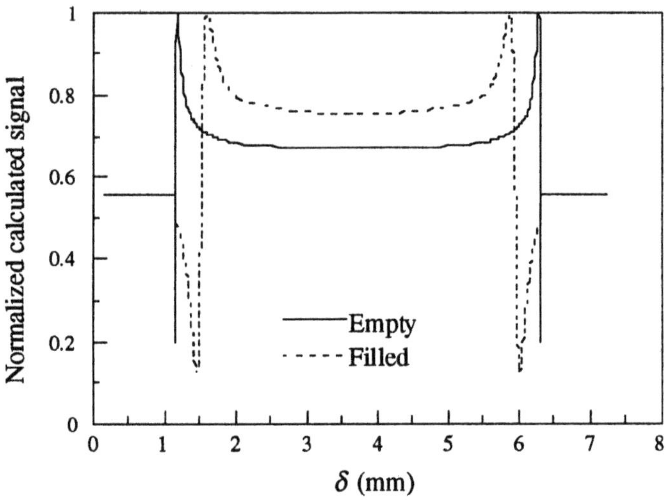

Fig. 4.11 Normalized calculated crack characteristic signals for an empty and filled (with rust powder) crack with width $W = 0.85$ mm and depth $D = 1.25$ mm at a frequency of 24 GHz (Huber *et al.*, 1997a).

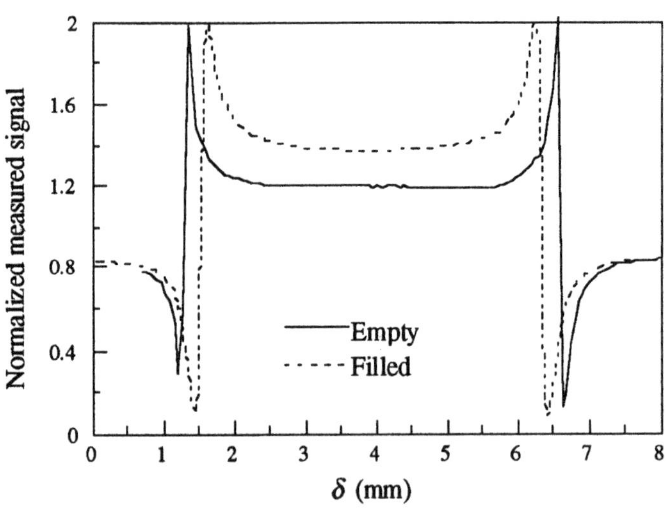

Fig. 4.12 Normalized measured crack characteristic signals for an empty and filled (with rust powder) crack with width $W = 0.85$ mm and depth $D = 1.25$ mm at a frequency of 24 GHz (Huber *et al.*, 1997a).

$$E_{ywi} = \sin\frac{\pi x}{a} e^{-j\gamma_w z} \tag{4.23a}$$

$$H_{xwi} = \frac{-1}{\eta_w}\sin\frac{\pi x}{a} e^{-j\gamma_w z} \tag{4.23b}$$

Here, the magnitude of the incident electric field is assumed to be unity, subscripts w and i denote waveguide and incident fields respectively and,

$$\gamma_w = \sqrt{k_o^2 - \left(\frac{\pi}{a}\right)^2} \tag{4.24a}$$

$$k_o = \frac{2\pi}{\lambda_o} \tag{4.24b}$$

$$\eta_w = \frac{k_o \eta_o}{\gamma_w} \tag{4.24c}$$

$$\eta_o = \sqrt{\frac{\mu_o}{\varepsilon_o}} \tag{4.24d}$$

where λ_o, k_o, ε_o and μ_o are the free-space wavelength, wavenumber, permittivity and permeability, respectively. η_o and η_w are the free-space and waveguide intrinsic impedances, respectively and γ_w is the propagation constant in the waveguide. Forcing the boundary conditions gives the following sought for electric and magnetic fields for the reflected higher-order TM modes inside the waveguide (subscript r denotes reflected fields):

$$E_{yr(TM)} = \sum_{m=1}^{\infty} A_{1m} \sin\frac{\pi x}{a}\cos\frac{m\pi y}{b} e^{j\gamma_{w(1m)}z} \tag{4.25a}$$

$$H_{yr(TM)} = \sum_{m=1}^{\infty} \frac{A_{1m}}{\eta_{w(1m)}} \sin\frac{\pi x}{a}\cos\frac{m\pi y}{b} e^{j\gamma_{w(1m)}z} \tag{4.25b}$$

$$E_{xr(TM)} = \sum_{m=1}^{\infty} \frac{b}{ma} A_{1m} \cos\frac{\pi x}{a}\sin\frac{m\pi y}{b} e^{j\gamma_{w(1m)}z} \tag{4.25c}$$

where A_{1m} ($m = 1,2,3,...$) are unknown coefficient to be determined and,

$$\gamma_{w(1m)} = \sqrt{k_o - \left(\frac{\pi}{a}\right)^2 - \left(\frac{m\pi}{b}\right)^2} \tag{4.26a}$$

$$\eta_{w(1m)} = \frac{\gamma_{w(1m)}\eta_o}{k_o} \tag{4.26b}$$

The electric and magnetic fields of the forward and reflected TM modes in the crack are given by:

$$E_{y(TM)} = \sum_{m=1}^{\infty}\left(B_{1m}e^{-j\gamma_{c(1m)}z} + C_{1m}e^{j\gamma_{c(1m)}z}\right)\cdot \sin\frac{\pi x}{a}\cos\frac{m\pi(y-c)}{W} \tag{4.27a}$$

$$H_{x(TM)} = \sum_{m=1}^{\infty}\left(-B_{1m}e^{-j\gamma_{c(1m)}z} + C_{1m}e^{j\gamma_{c(1m)}z}\right)\cdot \frac{1}{\eta_{c(1m)}}\sin\frac{\pi x}{a}\cos\frac{m\pi(y-c)}{W} \tag{4.27b}$$

$$E_{x(TM)} = \sum_{m=1}^{\infty}\left(B_{1m}e^{-j\gamma_{c(1m)}z} + C_{1m}e^{j\gamma_{c(1m)}z}\right)\cdot \frac{W}{am}\cos\frac{\pi x}{a}\sin\frac{m\pi(y-c)}{W} \tag{4.27c}$$

where B_{1m} and C_{1m} ($m = 1,2,3,...$) are coefficients to be determined and c (in this section) denotes the crack edge location in the waveguide aperture (Yeh and Zoughi, 1994a Fig. 4) and,

$$\gamma_{c(1m)} = \sqrt{k_o^2 - \left(\frac{\pi}{a}\right)^2 - \left(\frac{m\pi}{W}\right)^2} = -j\alpha \tag{4.28a}$$

$$\eta_{c(1m)} = \frac{\gamma_{c(1m)}\eta_o}{k_o} \tag{4.28b}$$

α represents the attenuation constant in the crack and all the unknown coefficients are subsequently found by forcing the boundary conditions (Yeh and Zoughi, 1994a; Yeh, 1995).

The reason for exploring the characteristics of these higher-order modes for fatigue/surface crack detection is that in the absence of a crack there are no higher-order modes present, hence theoretically the magnitude of the measured signal associated with higher-order modes is zero (Yeh, Ranu and Zoughi, 1994). However, the presence of a crack generates higher-order modes, and subsequently a finite amount of signal is measured. Hence, theoretically a signal-to-noise ratio (SNR) of equal to infinite is achieved which results in very sensitive (to the presence of a crack) measurements. In practice, however, the noise characteristics of the measurement system dictates the noise level. This high measurement sensitivity renders small fatigue/surface cracks (i.e. at their early stages of development) easily detectable. Moreover, it implies that for a given range of crack widths or opening lower microwave frequencies may be used with relatively high detection sensitivity, as will be seen later.

4.4.1 Results

Figure 4.13 shows the apparatus used for measuring the x-component of the overall electric field of the reflected higher-order modes, $E_x(z)$. The oscillator produces the microwave signal at a desired frequency which is then fed into a waveguide operating in its dominant mode. $E_x(z)$ is then measured (via a crystal detector) by a small probe located at a distance ℓ near the waveguide aperture. The measured voltage, which is proportional to the power associated with the higher-order modes (i.e. crystal detector operating in its square law region), is then recorded as a function of the scanning distance, δ. Figure 4.14 shows a close-up geometry of the relative location of the probe and the waveguide aperture. $\mathcal{T}$ is the distance between the probe and the waveguide wall, and as will be seen later the proper choice of this parameter can substantially increase the crack detection capability of this technique.

Higher-order TM modes are evanescent and their magnitudes decay exponentially with the distance from the waveguide aperture, ℓ (Pozar, 1990). However, close to the aperture their presence may be detected by sensing and measuring $|E_x(z)|$. Figure 4.16 shows $|E_x(z)|^2$ as a function of ℓ for a crack with a width of 0.5 mm, a depth of 0.5 mm and a length of 10.67 mm at 24 GHz when the crack is at the edge of the waveguide. $|E_x(z)|^2$ represents the square law property of the crystal diode detector which is used in the actual measurements. The exponential decay of $|E_x(z)|^2$ as a function of ℓ is evident, and as ℓ increases beyond 1 mm the signal level substantially decreases.

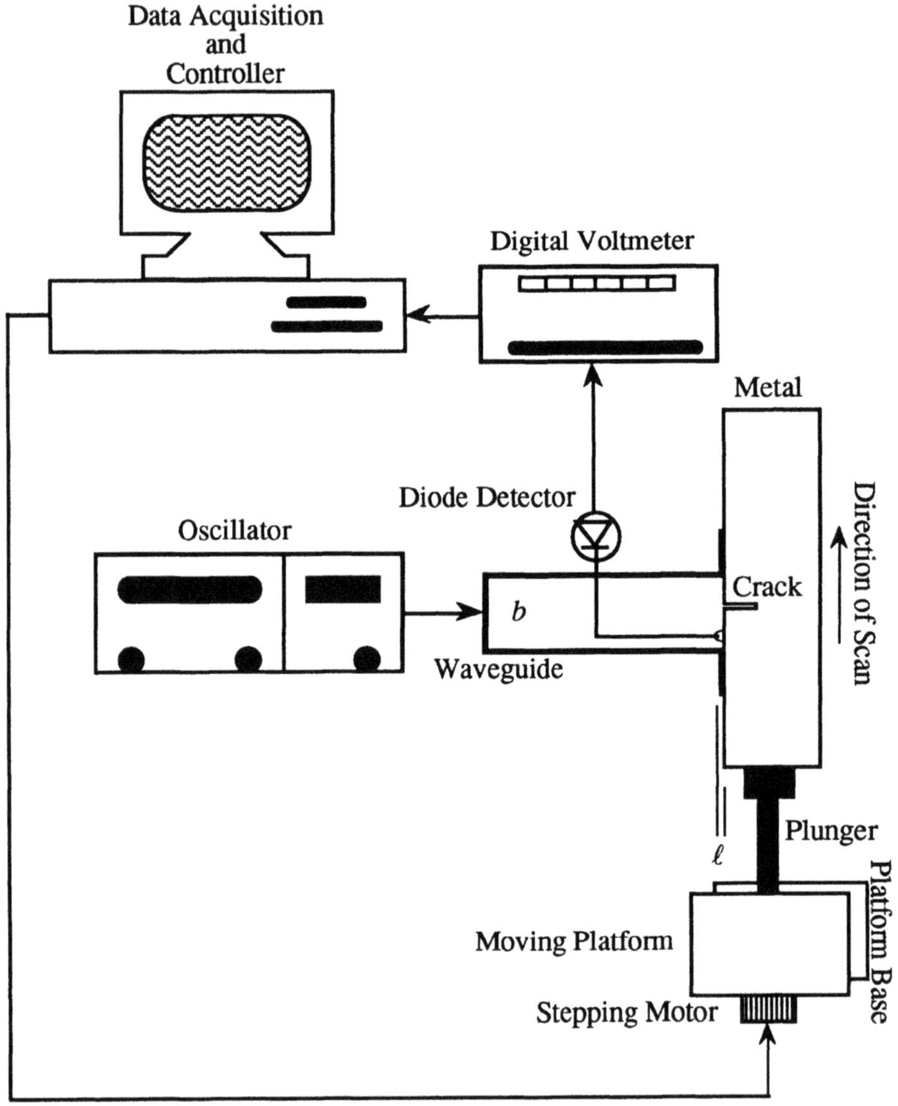

Fig. 4.13 Higher-order mode measurement apparatus.

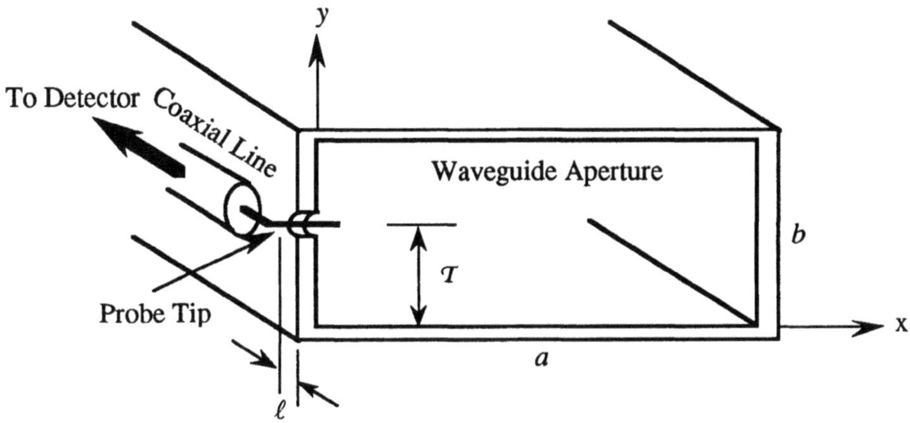

Fig. 4.14 The close-up geometry of the higher-order mode sensor showing the relative location of the probe and the waveguide aperture (Yeh, Ranu and Zoughi, 1994).

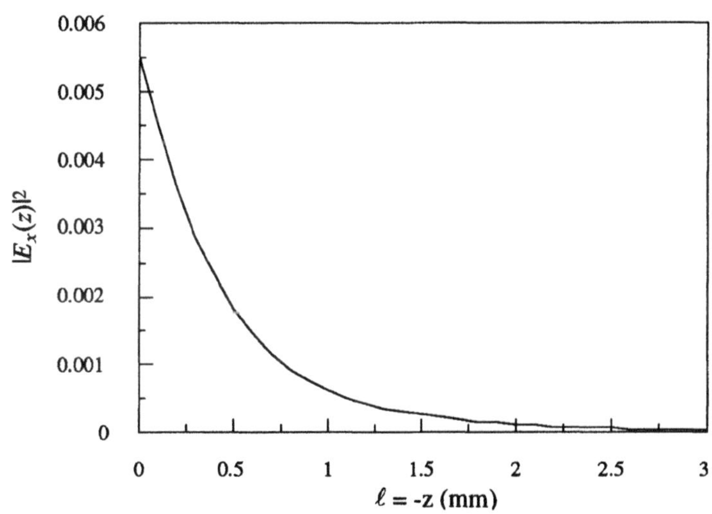

Fig. 4.15 $|E_x(z)|^2$ as a function of ℓ for a crack with width W = 0.5 mm, depth D = 0.5 mm and length L > 10.67 mm at 24 GHz (Yeh, Ranu and Zoughi, 1994).

To examine the optimal probe location for detecting the maximum magnitude of $E_x(z)$, the **higher-order mode crack characteristic signal** for two cracks with equal widths of 0.5 mm, equal lengths of 10.67 mm and respective depths of 1 mm and 1.5 mm were theoretically obtained. Figure 4.16 shows $|E_x(z)|^2$ as a function of $\xi = (T/b)$ at 24 GHz. The results show that maximum signal level occurs at $\xi \approx 0.17$. The optimal probe location for any given crack size turns out to be a function of the waveguide dimension, b, and the operating frequency and is approximately in the range of $0.11 \le \xi \le 0.20$ for the frequency range of 12.4-40 GHz (Yeh, 1994).

Figure 4.17 shows the theoretical variation of $|E_x(z)|^2$ as a function of the scanning distance, δ, for three different values of ξ for a crack with a width of 0.144 mm, a depth of 1.2 mm and a length of 22.86 mm at 12.4 GHz. When the probe is in the middle of the waveguide (long dashed line) two equal peak signal amplitudes are seen as the crack enters and exits the waveguide. When the probe location is at a distance $b/3$ from the broad wall of the waveguide (short dashed line) two unequal peak signal amplitudes are produced. The unequal peak signal amplitudes are due to the fact that when the crack enters the waveguide aperture it is closer to the probe than when it exits the waveguide aperture. Hence, the probe is influenced differently by the generated higher-order TM modes.

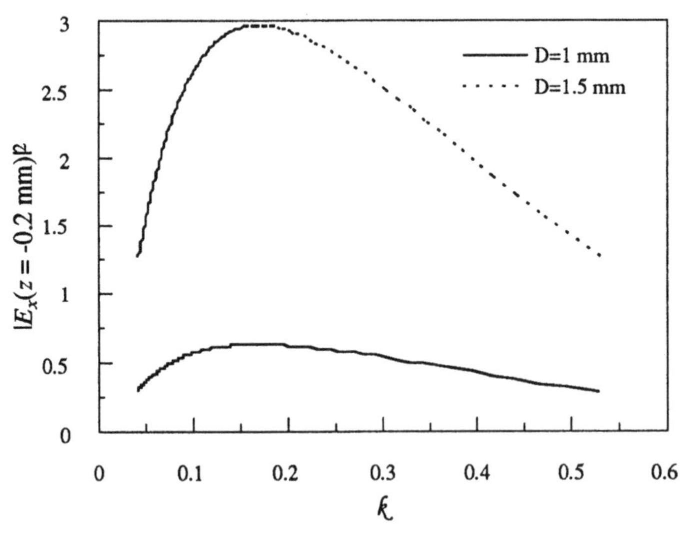

Fig. 4.16 Theoretical optimal probe location, ($\xi = T/b$), at 24 GHz for two cracks with equal widths W = 0.5 mm, equal lengths L = 10.67 mm and different depths (Yeh, Ranu and Zoughi, 1994).

Finally, at the optimal probe location (solid line) the same phenomenon, as in the previous case, occurs. However, this time due to the optimal choice of the probe location, the higher of the two unequal peak signal amplitudes is at a higher level than its counterpart in the previous case. Clearly, the correct choice of probe location increases the crack detection capability.

Figures 4.18 and 4.19 show the experimental results for the crack mentioned in Fig. 4.17 for two probe locations of $k_c = 0.5$ and $k_c = 0.17$, respectively. These results closely match, both in shape and relative magnitude, with the theoretical findings shown in Fig. 4.17. The results shown in Figs. 4.17-4.19 were calculated and measured at $\ell = 0.2$ mm.

It has been shown (Zoughi *et al.*, 1995) that increasing the operating frequency tends to enhance the detection of smaller (thinner or tighter) cracks. Hence, to demonstrate the capability of this technique to detect very small cracks, a standard fatigue specimen was subjected to cyclical loading until a fatigue crack was generated, as shown in Fig. 4.20. The width of this crack was measured, by a microscope, and varied between 1.9 micrometers and 4.9 micrometers. The crack was a through crack. This crack was scanned by a higher-order mode probe operating at 38 GHz. Fig. 4.21 shows the results of this experiment.

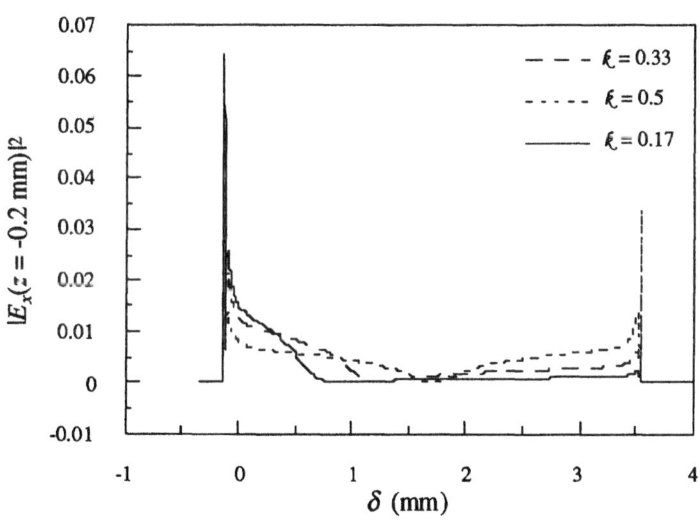

Fig. 4.17 $|E_x(z)|^2$ at z = -0.2 mm as a function of three different probe locations for a crack with width $W = 0.144$ mm, depth $D = 1.2$ mm and length $L = 22.86$ mm at 12.4 GHz (Yeh, Ranu, Zoughi, 1994).

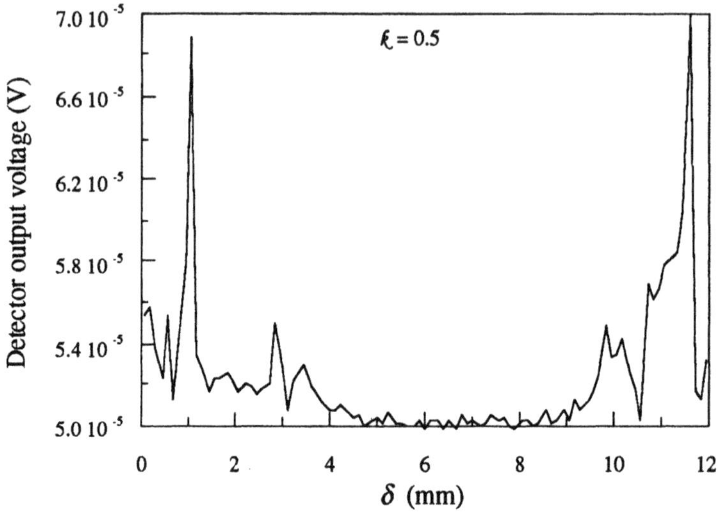

Fig. 4.18 Experimental results for a crack with width W = 0.144 mm, depth D = 1.2 mm and length L = 22.86 at 12.4 GHz for k = 0.5 (Yeh, Ranu and Zoughi, 1994).

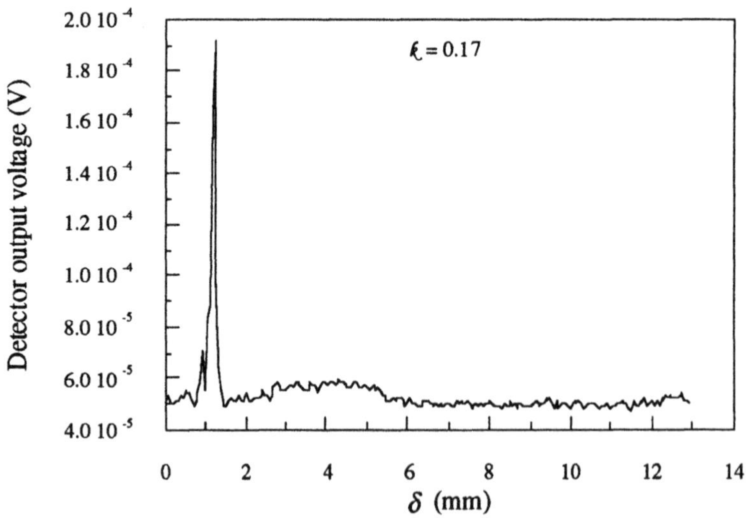

Fig. 4.19 Experimental results for a crack with width W = 0.144 mm, depth D = 1.2 mm and length L = 22.86 at 12.4 GHz for k = 0.17 (Yeh, Ranu and Zoughi, 1994).

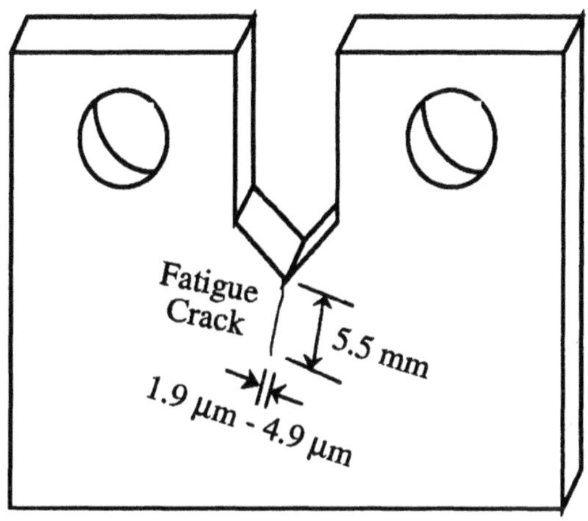

Fig. 4.20 Schematic of a standard fatigue specimen (Yeh, Ranu, Zoughi, 1994).

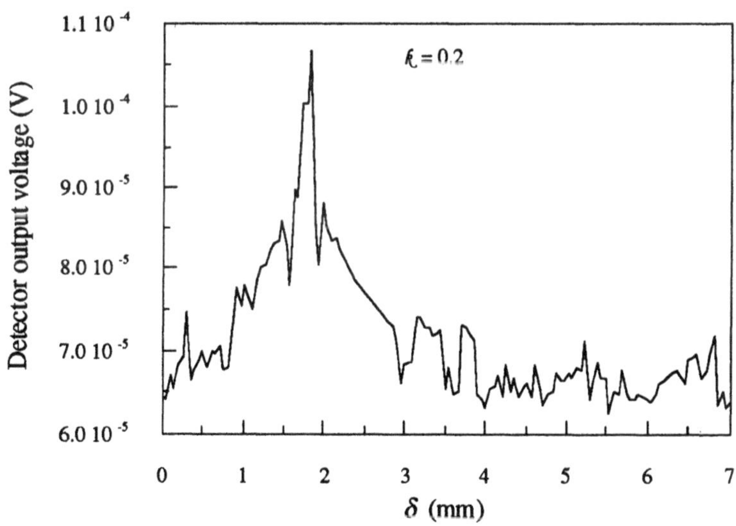

Fig. 4.21 Experimental result for the fatigue crack shown in Fig. 20 at 38 GHz for $\xi = 0.2$ (Yeh, Ranu, Zoughi, 1994).

The results show the capability of this technique to detect this fatigue crack (this experiment was repeated many times and each time the same results were obtained). The experiment was conducted in such a way as to ensure that the notch associated with the fatigue specimen was not exposed to the waveguide aperture during the scanning.

From Fig. 4.21 a SNR ≅ 3 dB is evident which is quite adequate for detection. However, it must be mentioned that the measurement system used in this experiment suffered from internal noise problem (≈65 microvolts). However, not using a computer controlled recording system, data was obtained with a more sensitive measurement system with noise characteristics of less than 6 microvolts. This translates to a SNR ≅ 13 dB. Another important measurement criterion is the resolution of the scanner which scans the waveguide over the metal surface (or vise versa). Figure 4.17 shows that the signal peak indicating the presence of the crack is very narrow in width (as a function of δ). Therefore, if the resolution of the scanner is coarse, the maximum peak amplitude may not be detected. The consequence of this is lower SNR or lower detection sensitivity. Conversely, the consequence of a scanner with fine resolution may be higher SNR or increased detection sensitivity.

There are several issues to be considered regarding the higher-order mode approach. Although the higher-order TM modes are assumed to be the significant set of modes for long cracks, depending on the width or opening of the crack other modes and field components (not just x-component of the electric fields associated with the higher-order TM modes) with lesser influence may also contribute to the process. From a practical point of view, the effect of probe in disturbing the fields at the waveguide aperture may be a significant issue. The probe should be long enough to pick up high enough higher-order mode signal amplitude necessary for detecting the crack. However, it should be short and thin enough so as to not to perturb the aperture fields significantly. This is similar to the design criterion for a slotted waveguide probe. Extensive measurements have been conduced to show the influence of the probe characteristics and the frequency of operation on crack detection using of the dominant and the higher-order mode probes in tandem, as outlined in (Zoughi *et al.*, 1995 Chapter VIII).

4.5 TYPICAL MEASUREMENT RESULTS

During 1994-1995 an extensive effort was undertaken to study the many features of this microwave crack detection and evaluation technique as a function of many parameters such as the crack width and depth, the frequency of operation, the incident power level and for empty, filled and covered cracks (Zoughi *et al.*, 1995). In this section some typical results obtained in this investigation are presented.

4.5.1 Influence of crack width and depth on characteristic signal

It has been shown that the detector voltage level, when the crack is inside the waveguide aperture, is a function of the crack width and more significantly of the crack depth. Similarly, the distance between the two sharp transitions in a crack characteristic signal is primarily a function of crack width (plus the narrow dimension of the waveguide aperture). Many measurements were conducted to check the degree of influence that the crack width and depth have on the crack characteristic signal. Figure 4.22 shows the normalized crack characteristic signals for a crack with a depth of 1 mm and various widths at 24 GHz. The normalization is with respect to the signal level when the crack is outside of the waveguide aperture (i.e. the short-circuited signal level). The results indicate that as the width changes, the distance between the two sharp transitions changes, as expected.

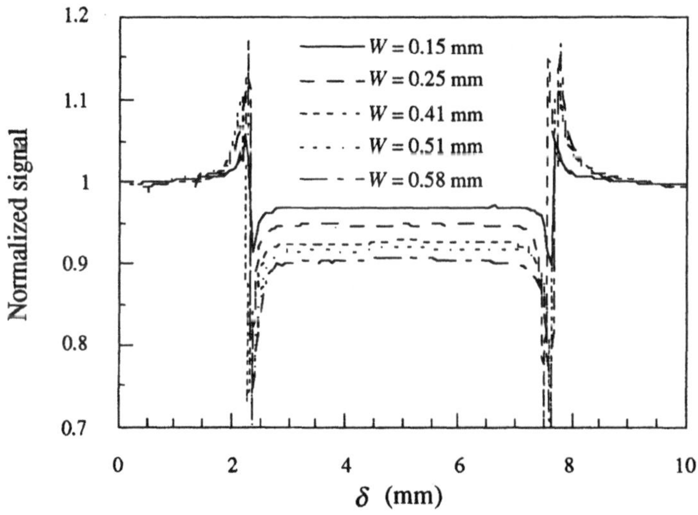

Fig. 4.22 Normalized crack characteristic signals for cracks with equal depths D = 1 mm and various widths at 24 GHz (Zoughi *et al.*, 1995).

However, the middle signal level changes as well. This is an indication of the fact that the signal level in the middle of the crack characteristic signal (i.e. crack in the middle of the waveguide) is also influenced by the crack width.

Figures 4.23-4.25 show the results of similar experiments except for cracks with equal widths of 0.51 mm and various depths at 24 GHz. The effect of depth variation on the signal level in the middle of the characteristic signals is much more than the effect of width. Furthermore, the distance between the two sharp transitions also changes as a function of depth, and perhaps more significantly than expected earlier.

Based on these measurements one may conclude that a change in the width of a crack (for fixed depth) does not alter the overall shape of the crack characteristic signal although the distance between the two sharp transitions and the signal level in the middle of the characteristic signal change (Fig. 4.22). When the width is fixed and the depth changes the shape of the crack characteristic signal changes as well (gradual at the outset and drastically as the depth increases). There is also another important phenomenon to note. For cracks with fixed widths and at fixed operating frequencies there may exist a certain range of depths for which the detected signals for when the crack is inside and outside the waveguide aperture are equal.

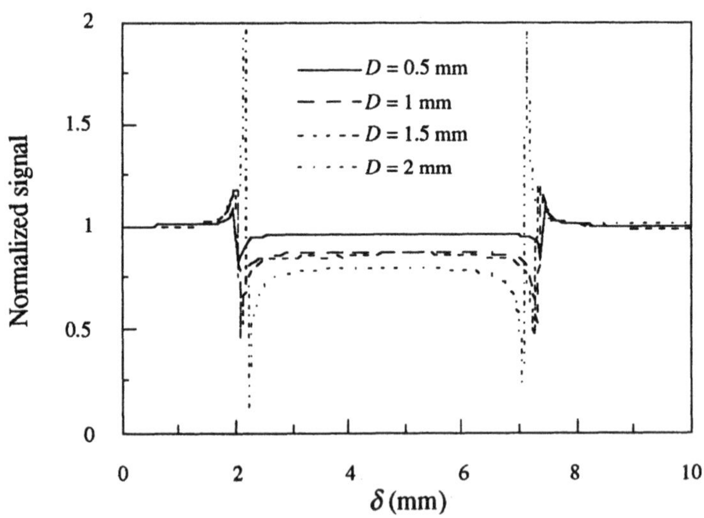

Fig. 4.23 Normalized crack characteristic signals for cracks with equal widths W = 0.51 mm and various depths at 24 GHz (Zoughi *et al.*, 1995).

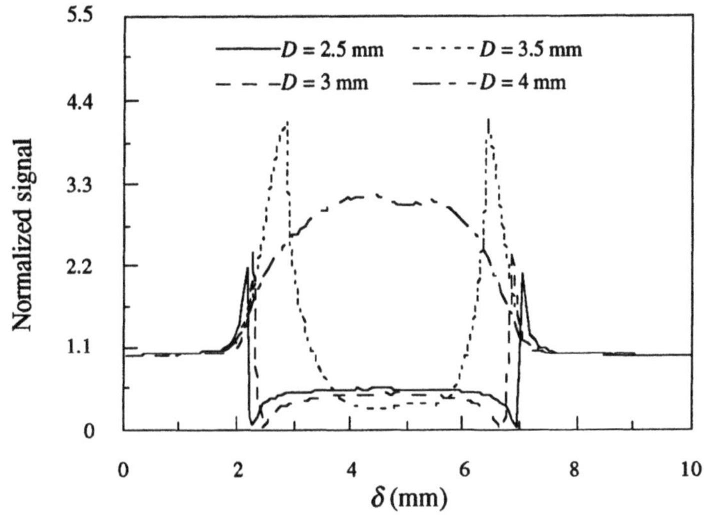

Fig. 4.24 Normalized crack characteristic signals for cracks with equal widths W = 0.51 mm and various depths at 24 GHz (Zoughi *et al.*, 1995).

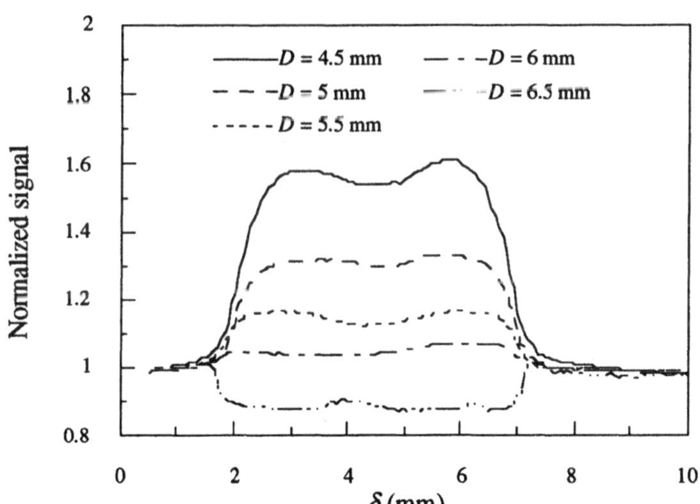

Fig. 4.25 Normalized crack characteristic signals for cracks with equal widths W = 0.51 mm and various depths at 24 GHz (Zoughi *et al.*, 1995).

This renders the crack undetectable (for example see Fig. 4.25, $D = 6$ mm). This may be avoided by using multiple detector diodes at different positions in the standing wave pattern or by a slight change in the operating frequency.

Similar measurements were conducted on steel specimens as well. Due to the difficulties associated with machining of these specimens only a few cracks were cut in steel. Figure 4.26 shows the results of changing depth on crack characteristic signal, at 24 GHz, and for a crack with a width of 0.20 mm. As expected the characteristic signals for these cracks have the same overall shape as those cracks in the aluminum plates. Furthermore, the influence of varying crack depth is similar to those described for cracks in aluminum plates.

4.5.2 Influence of detector location on characteristic signal

The diode detector that probes the standing wave properties inside the waveguide is usually located exactly in the middle of the standing wave maximum and minimum. However, for some crack dimensions this location may be changed for increased detection sensitivity.

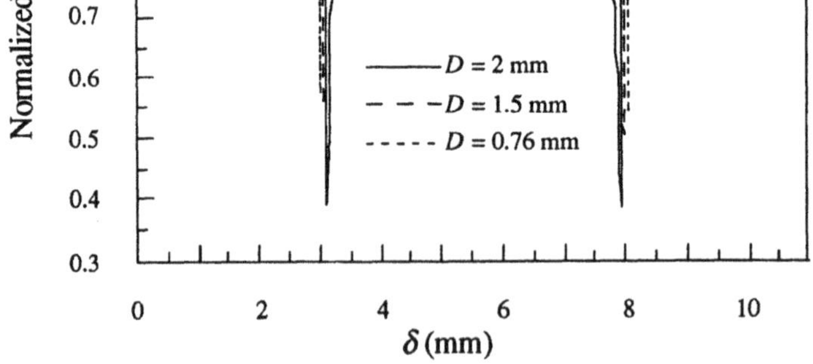

Fig. 4.26 Normalized crack characteristic signals for cracks, in steel plate, with equal width $W = 0.20$ mm and various depths at 24 GHz (Zoughi *et al.*, 1995).

Figure 4.27 shows a sample set of normalized crack characteristic signals for a crack with a width of 0.85 mm and a depth of 0.93 mm at 10.5 GHz. Each characteristic signal in this set is obtained for a different diode location within the waveguide (arbitrarily chosen locations such as 97.5 mm away from the waveguide aperture). Besides the fact that all of these curves have similar overall shape, it must be noted that the normalized level of the signal in the middle of each characteristic signal varies as a function of diode location. This confirms the notion that depending on the location of the diode detector more or less sensitivity to the presence of the crack may be obtained. Figure 4.28 shows the first 3 mm of the scanning distance of the normalized crack characteristic signals shown in Fig. 4.27. One can clearly see that for some diode locations the difference between the short circuit level (crack outside the waveguide) and when the crack is totally inside the waveguide aperture is negligible (i.e. for 93 mm which is very near a null of the standing wave pattern). If a coarse resolution scanner is used in such cases the crack may not at all be detected. There is also a reversal of transition as a function of diode location. This indicates that these diode locations are on the other side of a minimum in the standing wave pattern. This is also the reason for the apparent difference, for example, between Figs. 4.10 and 4.22.

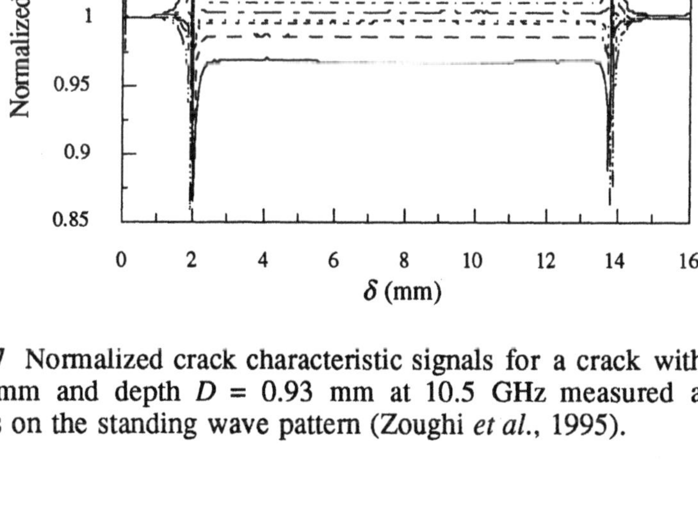

Fig. 4.27 Normalized crack characteristic signals for a crack with width W = 0.85 mm and depth D = 0.93 mm at 10.5 GHz measured at various locations on the standing wave pattern (Zoughi *et al.*, 1995).

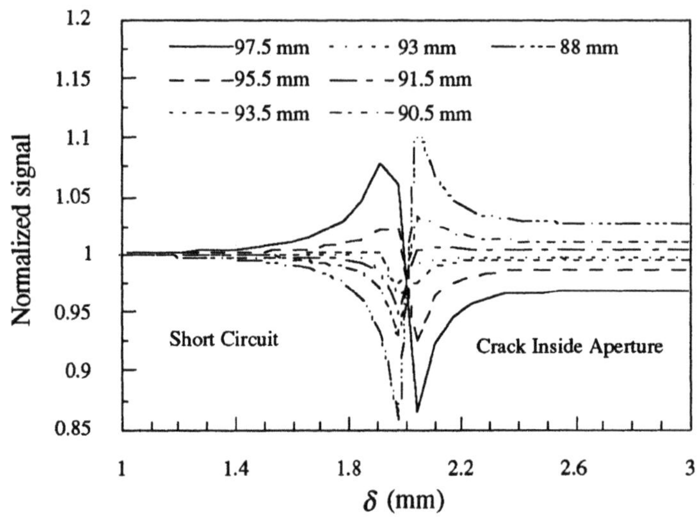

Fig. 4.28 A portion of the normalized crack characteristic signals for a crack with width W = 0.85 mm and depth D = 0.93 mm at 10.5 GHz measured at various locations on the standing wave pattern, as shown in Fig. 4.27 (Zoughi *et al.*, 1995).

4.5.3 Influence of frequency on characteristic signal

A waveguide can operate in its dominant mode in a fixed range of frequencies which is known as the waveguide bandwidth. Thus, if one uses a waveguide probe operating in Ku-band (12-18 GHz), it is possible to use any frequency in this range as the operating frequency of the probe. Subsequently, it is important to investigate the effect of the operating frequency given a waveguide probe.

It must be noted that the effect of changing frequency may be viewed as an electrical change in the dimensions of the crack for a frequency (i.e. changing the ratio between the operating wavelength and the dimensions of the crack). Thus, any change in the operating frequency is expected to have similar influences on the properties of a crack characteristic signal as the change in crack dimensions would have (i.e. change in the signal level in the middle of a crack characteristic signal and widening or narrowing of the distance between the two sharp peaks). Furthermore, a change in the operating frequency causes a shift in the standing wave pattern. When a diode detector is located at a fixed distance away from the waveguide aperture, and the frequency is changed the diode will no longer be located in the same relative position in the standing wave pattern (i.e. electrical distances have changed). Therefore, for the experiments conducted in this section slotted waveguides were used which allow for positioning of the diode detector anywhere in the standing wave pattern. Thus, after changing the operating frequency the diode detector was placed in the middle of the standing wave voltage for all measurements in this section. In this way the measurements conducted at various frequencies may be compared.

Figures 4.29 shows the normalized crack characteristic signals at three frequencies in Ku-band for a crack with a width of 0.20 mm and a depth of 2.5 mm. As expected, the changes in the crack characteristic signals are similar to a change in crack dimensions. In addition to the expected changes in the characteristics signals, another very important practical point may be made. For this crack at 14 GHz the changes in the crack characteristic signal are not very abrupt. However, at 18 GHz, in the same band, the changes are much more abrupt. This translates into a better detection of this crack at 18 GHz than at 14 GHz. The results clearly indicate that the operating frequency is an important parameter for increasing crack detection sensitivity. The results of the measurements on exposed cracks, using the dominant mode detection method, are indeed very positive and encouraging.

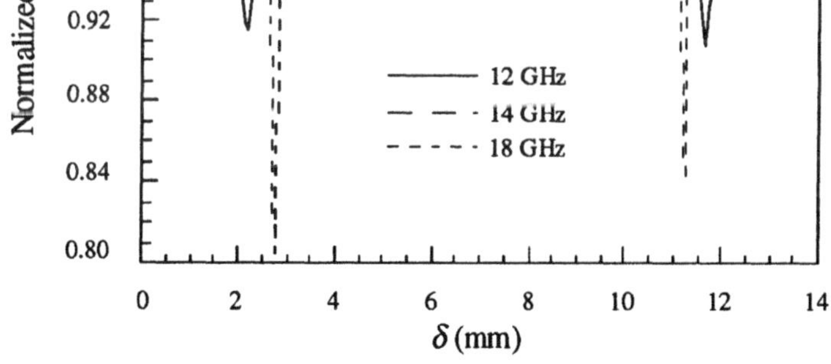

Fig. 4.29 Normalized crack characteristic signals for a crack with width $W = 0.2$ mm and depth $D = 2.5$ mm at three different frequencies (Zoughi *et al.*, 1995).

4.5.4 Filled cracks

Rust, dirt, chemicals and other substances may fill a crack at any time, particularly if the crack is old. Additionally, any potential distinction between a filled and non-filled crack may render information about when the crack may have been generated (i.e. recent or old). Consider a crack to be a microwave cavity fed by a waveguide. This is another modeling approach to the problem of a crack being scanned by a rectangular waveguide aperture. When the crack is empty its cavity related characteristics such as its quality factor and its resonant frequency are set by its dimensions, the frequency of operation and the properties of the feeding structure. However, when the cavity is filled with a dielectric material its quality factor and resonant frequency will be different than when it is empty. The dielectric properties of the material filling the cavity influences these changes (permittivity affects the resonant frequency and loss factor influences the quality factor). Therefore, the crack characteristic signal for an empty crack is expected to be different than when the crack is filled with a dielectric material. Thus, it is expected that the distance between the two sharp transitions and the signal level when the crack is in the middle portion of the waveguide aperture to be different for filled as opposed to empty cracks. The latter change may also be considered to be as a result of the fact that a filled crack electrically looks deeper than the same crack when empty (i.e. larger electrical depth). As has been established thus far, a change in the depth of a crack significantly influences this middle signal level in addition to less significant influence on the distance between the two sharp transitions. Filled crack experiments were conducted using chemically produced rust powder (Fe_2O_3) as the crack filler.

Figures 4.30-4.32 show the normalized crack characteristic signals, recorded at 24 GHz, for empty and filled with rust powder cracks/slots, all with a width of 0.51 mm and depths of 1 mm, 2.5 mm and 3 mm, respectively. These results primarily show a progressive change in the middle signal level, as expected. Additionally, Figs. 4.31 and 4.32 not only show this change, they also indicate the somewhat abrupt change in the overall shape of the characteristic signal. This change in the shape is very similar to the change between D = 4.5 mm and D = 5.5 mm for empty cracks (Fig. 4.25 in section 4.5.1). This once again confirms that filled cracks with electrically larger depths look similar to deep cracks.

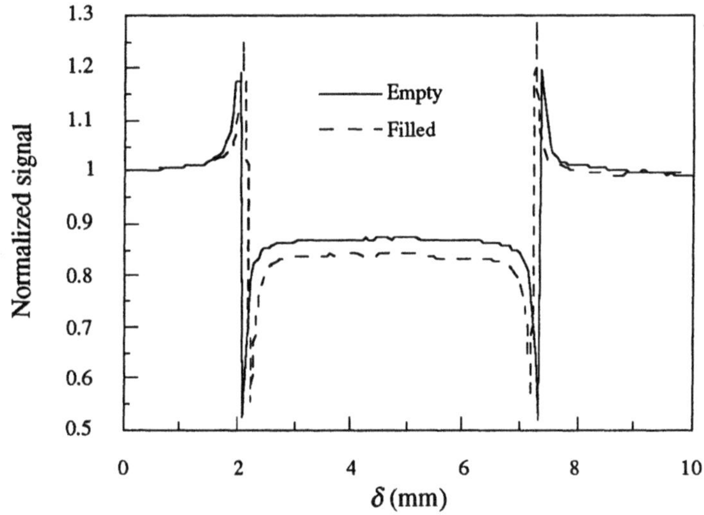

Fig. 4.30 Normalized crack characteristic signals for an empty and a filled crack (with Fe_2O_3) with width W = 0.51 mm and depth D = 1 mm at 24 GHz (Zoughi *et al.*, 1995).

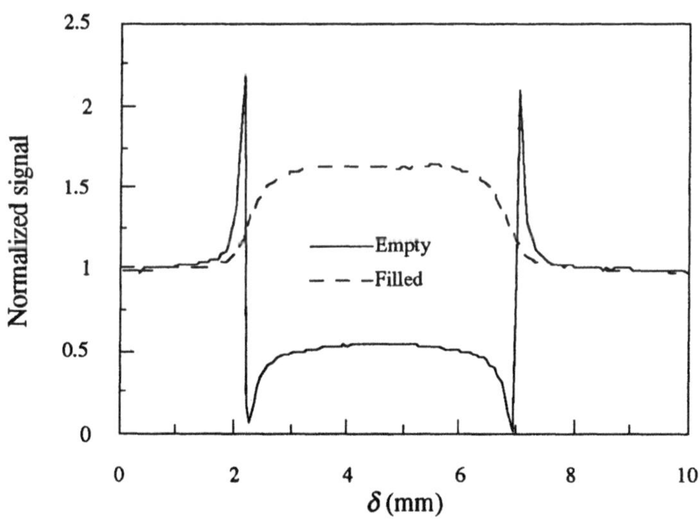

Fig. 4.31 Normalized crack characteristic signals for an empty and a filled crack (with Fe_2O_3) with width W = 0.51 mm and depth D = 2.5 mm at 24 GHz (Zoughi *et al.*, 1995).

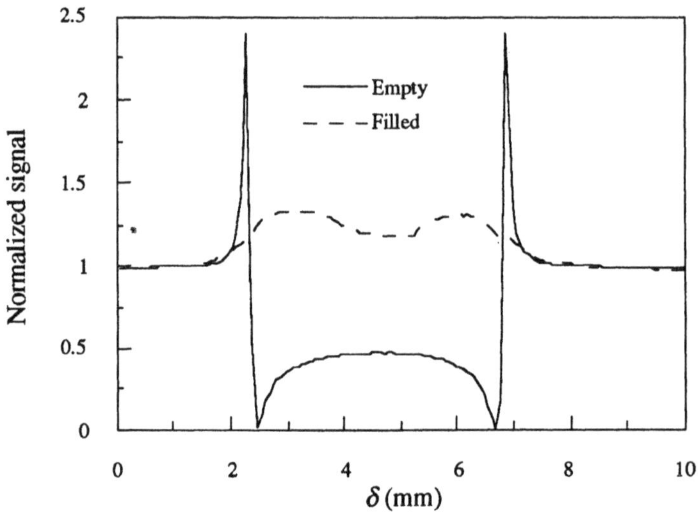

Fig. 4.32 Normalized crack characteristic signals for an empty and a filled crack (with Fe_2O_3) with width W = 0.51 mm and depth D = 3 mm at 24 GHz (Zoughi *et al.*, 1995).

Figures 4.33-4.36 show the normalized crack characteristic signals, recorded at 24 GHz, for slots filled with rust powder all with a width of 0.51 mm and depths of 3.5 mm, 4.5 mm, 5.5 mm and 6.5 mm, respectively. The most important feature associated with these results is that for some depths there is hardly any change in the characteristic signal (i.e. D = 3.5 mm for filled crack and D = 6.5 mm for empty crack) rendering the crack difficult (if not impossible) to detect. Furthermore, as the depths of the cracks increase the overall shape of the characteristic signal changes significantly. As expected, this feature was shown to be a function of crack depth. Referring to Figs. 4.33 and 4.34, when the crack is filled its detection is difficult. However, as the depth increases beyond 4.5 mm (Figs. 4.35 and 4.36) the filled normalized crack characteristic signals progressively take their original overall shape (rendering the crack detectable). The same consideration also applies to the empty crack with a depth of 6 mm (i.e. not easily detectable).

For comparison purposes, the results obtained for filled cracks, for depths to 4 mm, are shown in Figs. 4.37 and 4.38. Specifically, the progressive alteration to the overall normalized crack characteristic signal is evident.

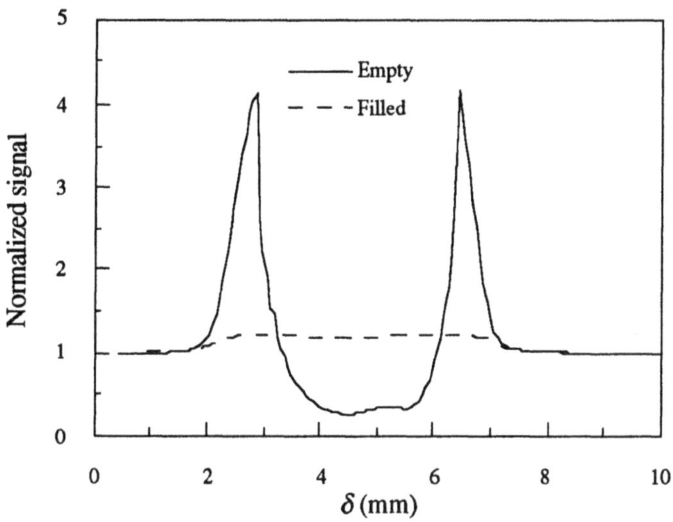

Fig. 4.33 Normalized crack characteristic signals for an empty and a filled crack (with Fe_2O_3) with width W = 0.51 mm and depth D = 3.5 mm at 24 GHz (Zoughi *et al.*, 1995).

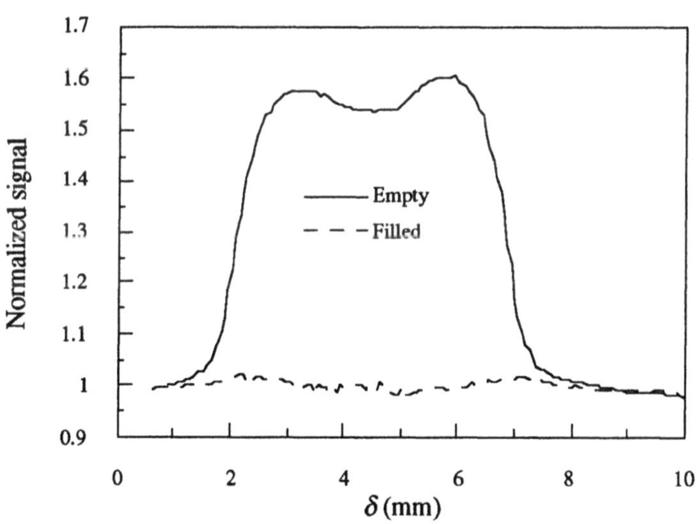

Fig. 4.34 Normalized crack characteristic signals for an empty and a filled crack (with Fe_2O_3) with width W = 0.51 mm and depth D = 4.5 mm at 24 GHz (Zoughi *et al.*, 1995).

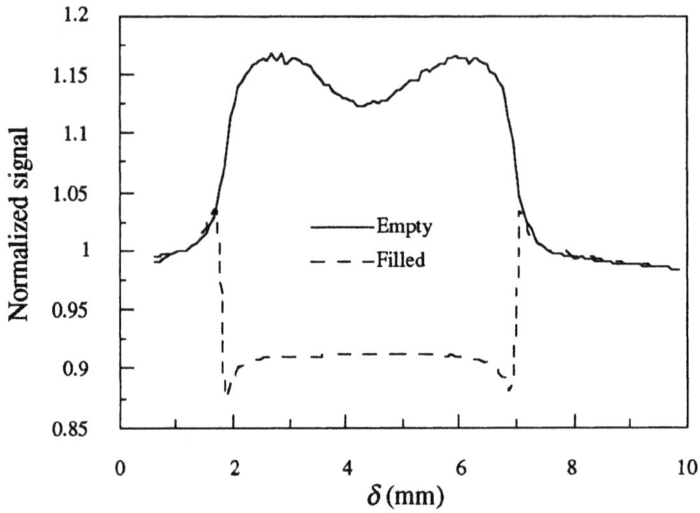

Fig. 4.35 Normalized crack characteristic signals for an empty and a filled crack (with Fe_2O_3) with width W = 0.51 mm and depth D = 5.5 mm at 24 GHz (Zoughi *et al.*, 1995).

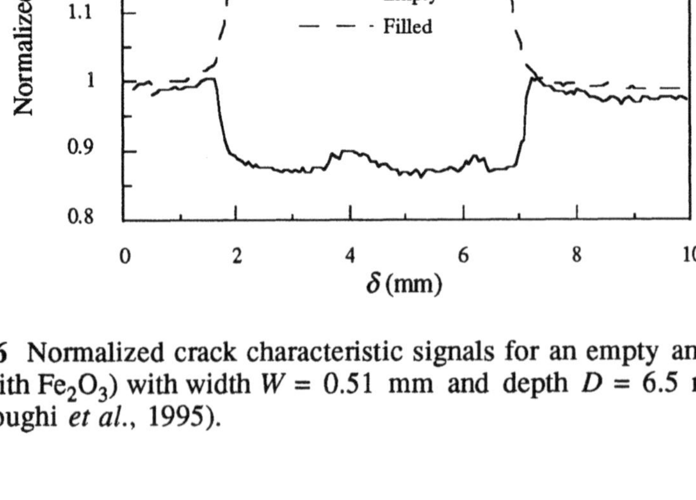

Fig. 4.36 Normalized crack characteristic signals for an empty and a filled crack (with Fe_2O_3) with width W = 0.51 mm and depth D = 6.5 mm at 24 GHz (Zoughi *et al.*, 1995).

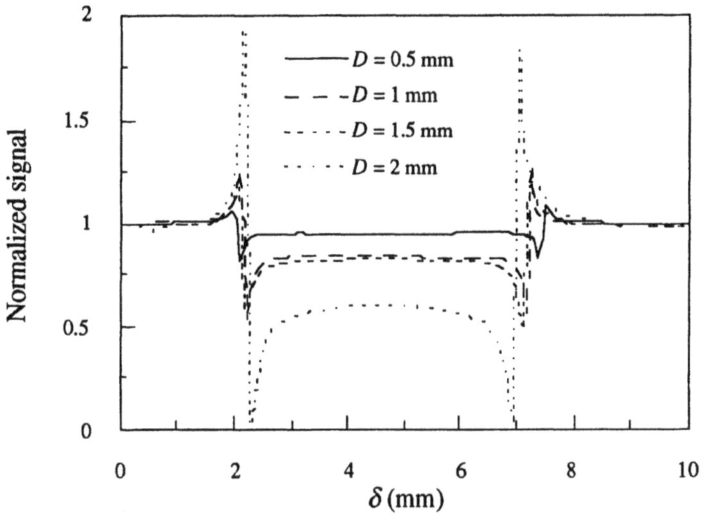

Fig. 4.37 Normalized crack characteristic signals for an empty and a filled crack (with Fe_2O_3) with width W = 0.51 mm and various depths at 24 GHz (Zoughi *et al.*, 1995).

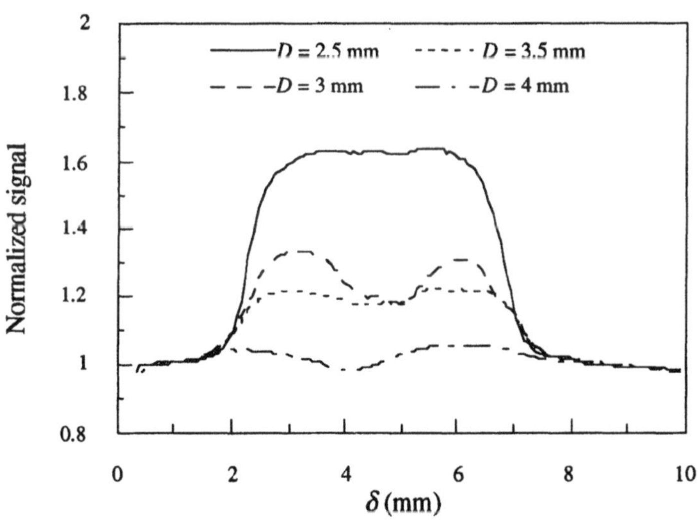

Fig. 4.38 Normalized crack characteristic signals for a filled crack (with Fe_2O_3) with width W = 0.51 mm and various depths at 24 GHz (Zoughi *et al.*, 1995).

The results shown in this section are very important as they indicate the utility of this microwave crack detection method in a complementary fashion with the dye penetrant method, since the dye filling a crack is a dielectric material. In all cases, when a crack characteristic signal does not render the crack detectable, a slight variation in the operating frequency or the use of multiple detector diodes, positioned at several locations on the standing wave pattern, will result in the successful detection of the crack. The same may be accomplished by placing a thin layer of dielectric sheet in front of the waveguide aperture (i.e. covered crack, as will be discussed later). In addition, the use of higher-order mode detection in tandem with the dominant mode detection will render all cracks detectable. Dominant mode and higher-order mode detection can be simultaneously accomplished with one waveguide probe.

Figures 4.39 and 4.40 show similar results, (only filled cases shown) for a crack with a depth of 2 mm and various widths. There is an interesting feature observed from these results. Thus far, all measurements have indicated that width variations tend not to change the overall shape of the normalized crack characteristic signal, but variation in depth does indeed change the overall shape. Figures 4.39 and 4.40 show that changing width also alters the overall shape of the crack characteristic signal for certain depths. It is perhaps more accurate to associate the operating wavelength to these types of general statements in such a way that some ratio of the width and depth (filled or empty) to the operating wavelength is considered.

In this section the influence of dielectric fillers such as dirt, rust, etc., on the measured characteristic signal was demonstrated. A filled crack has a longer electrical depth than an empty crack depending on the dielectric properties of the filling material. Thus, the effect of dielectric fillers should resemble a deeper crack. The experimental results confirmed this fact to be true. Chemically produced rust powder was used as fillers in this study. Results of another filler (beeswax) are shown in section 4.7 as part of filled crack tip location investigation.

4.5.5 Covered cracks

One of the most important advantages of this microwave methodology is evident when it is used for detecting surface cracks under dielectric coatings. Since microwave signals easily penetrate inside dielectric materials, this methodology should be able to detect cracks under dielectric coatings of various thicknesses. It must be noted that dielectric coatings such as paint, corrosion preventative substances, etc., may have various thicknesses although they are generally not very thick.

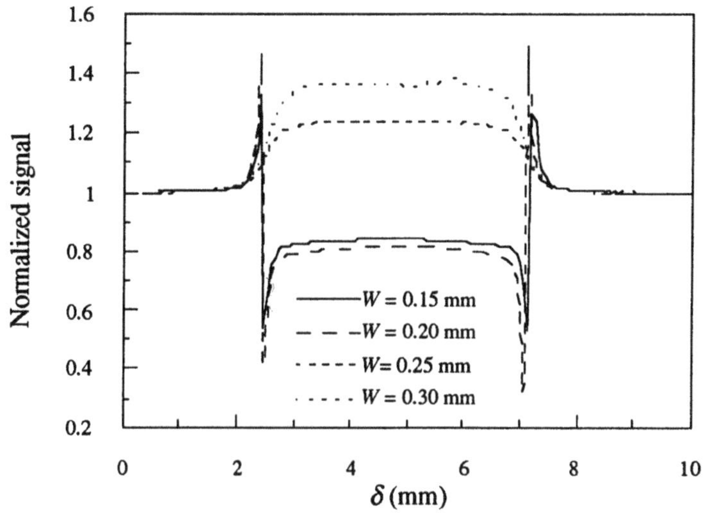

Fig. 4.39 Normalized crack characteristic signals for filled cracks (with Fe_2O_3) with depth $D = 2$ mm and various widths at 24 GHz (Zoughi *et al.*, 1995).

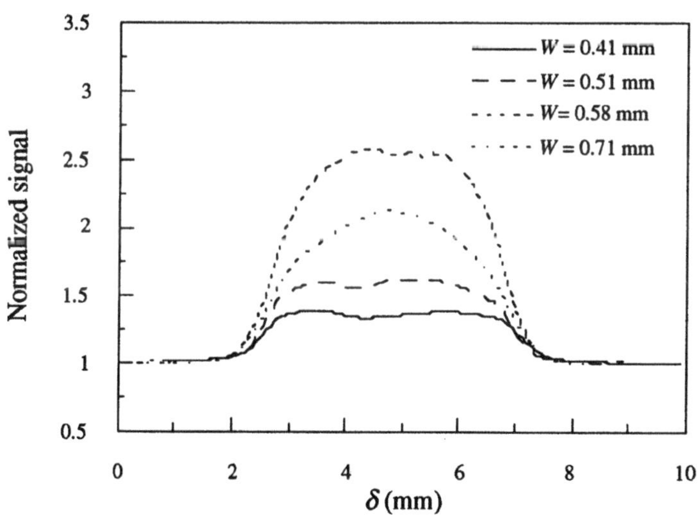

Fig. 4.40 Normalized crack characteristic signals for filled cracks (with Fe_2O_3) with depth $D = 2$ mm and various widths at 24 GHz (Zoughi *et al.*, 1995).

Therefore, it is important to establish the potentials and limitations of this microwave approach as a function of the dielectric coating thickness covering a crack. In this section the results of several measurements, at 24 GHz, on cracks covered with thin sheets of wrapping paper (simulating paint with different thicknesses) are discussed. Wrapping paper has similar dielectric properties as common paint. Furthermore, wrapping paper (with uniform thickness of 0.04 mm) may easily be stacked on top of each other to simulate various coating thicknesses. The results of several measurements on steel samples are also discussed. Additionally, the effect of increased input power on detecting cracks under thicker coatings was also investigated, and very promising results were obtained.

The results of the previous sections were normalized with respect to the short circuit signal level (i.e. when the crack is outside the waveguide aperture). In the case of covered cracks, the short circuit level changes as a function of the coating thickness. Thus, for the most part in this section, the recorded crack characteristic signals are shown and not their normalized versions.

A slot with a width of 0.51 mm and a depth of 2 mm was covered with up to twenty six sheets of wrapping paper (coating thickness of 0-1.04 mm) and their crack characteristic signals were recorded at 24 GHz, as shown in Figs. 4.41-4.44. As the coating thickness increases the overall level of the recorded signal decreases. This is primarily due to the fact that the standing wave pattern inside the waveguide is very different when it is terminated by a conductor (i.e. short circuit) and when it is terminated by a dielectric covered conductor. The difference between the maximum and the minimum levels of the standing wave decreases (i.e. decreasing SWR). Furthermore, there will be a shift in the standing wave location compared to the case of no coating which in turn puts the diode location at a different position on the standing wave pattern. The incident signal is also attenuated by the two-way travel in the coating (although this attenuation may be small in the case of low-loss dielectric coatings), contributing to a decreased signal level when detecting covered cracks. The signal level in the middle of the crack characteristic signal decreases as a function of increasing coating thickness, however so does the signal level when the crack is outside of the waveguide aperture. This is true for the first few layers of the wrapping paper, however the difference between these two levels increases as the thickness increases beyond three or four layers. The distance between the two sharp transitions increases also as a function of the increasing coating thickness. This is due to the fact that the incident microwave signal is able to interact with the crack edges (i.e. the aperture "sees" the crack edge or radiation of the microwave signal into the coating) before it comes over it, unlike the flush case. Furthermore, the sharpness associated with the transitions degrades rapidly as a function of

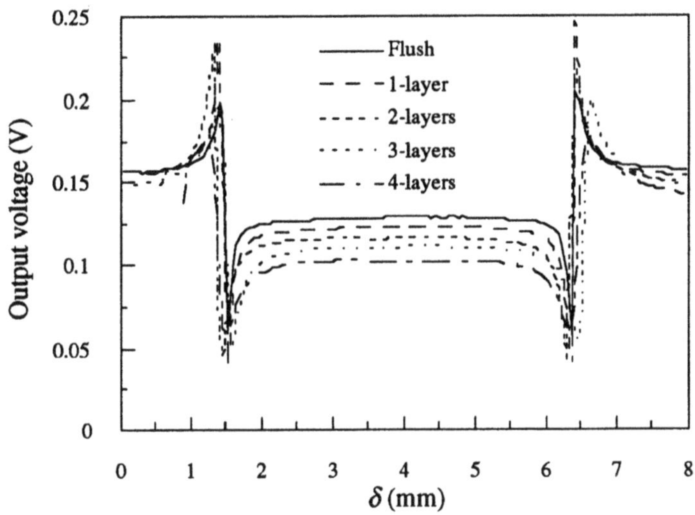

Fig. 4.41 Crack characteristic signals for a crack with width W = 0.51 mm and depth D = 2 mm at flush and when covered with four different layers of wrapping paper (coating thickness range of 0.04-0.16 mm) at 24 GHz (Zoughi *et al.*, 1995).

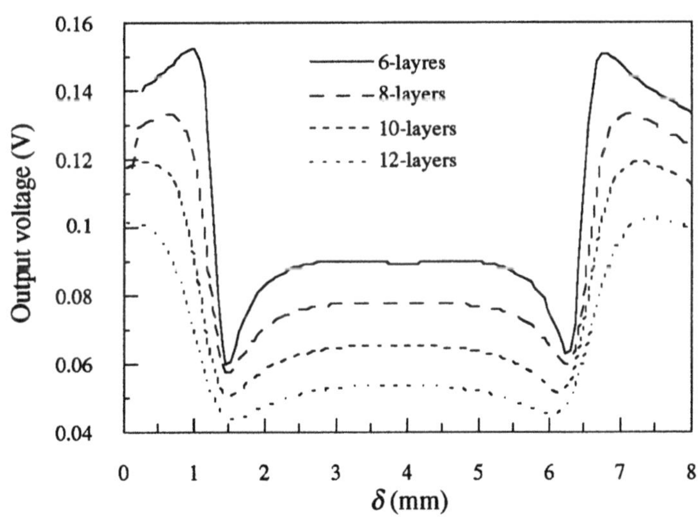

Fig. 4.42 Crack characteristic signals for a crack with width W = 0.51 mm and depth D = 2 mm at flush and when covered with four different layers of wrapping paper (coating thickness range of 0.24-0.48 mm) at 24 GHz (Zoughi *et al.*, 1995).

the increasing coating thickness. This is an important feature associated with covered cracks. This means that in practice one may not need a fine resolution scanner for detecting covered cracks, since the possibility of jumping over the transitions is low, particularly for cracks whose depth is such that the middle signal level and the signal level when the crack is outside the aperture are almost equal. Even for exposed cracks it may be advantageous to use a dielectric coating in front of the waveguide aperture to decrease the possibility of missing the sharp transitions in the crack characteristic signal. Another important observation is that even with twenty six sheets of wrapping paper the crack is still detectable. Actually the results indicate that the crack may be covered with additional sheets before it becomes undetectable. The dynamic range of the detected voltage for these crack characteristic signals is quite large indicating that many more layers may cover this crack, and it will still be detectable. Coating thicknesses of greater than a couple of millimeters is considered unusually thick for paint and corrosion preventative substances. However, the results indicate the cracks under such thick coatings may also be detected using this microwave method.

As the thickness of the coating increases, the crack characteristic signal shows less variation in its overall shape making the detection of the crack more difficult. This is also true as the loss factor of the dielectric coating increases. Since the standing wave inside the waveguide is formed as a result of the interference between the reflected by the cracked metal specimen and the incident wave, it is reasonable to assume that an increase in the incident signal power will increase the reflected signal power. Consequently, cracks covered with thicker coatings may be detected. To demonstrate this, a crack with a width of 0.2 mm and a depth of 1.5 mm, milled on a steel specimen, and covered with three different thicknesses of wrapping paper was scanned. A frequency of 17 GHz was experimentally found to be optimum for these measurement. The crack characteristic signals were recorded for the input power from 0 dBm (1 milliwatt) to 12 dBm at 4 dBm intervals. dBm is a unit of power in decibels calculated as (10log) of power in milliwatts. Figures 4.45 and 4.46 show the results of these measurements. From these figures it is evident that when the input power is low (0 dBm) the crack characteristic signal does not exhibit enough variations in its properties, hence rendering the crack difficult to detect. However, as the input power is increased, the crack characteristic signal begins to show some variations in its overall shape making the crack easily detectable. For 12 dBm the crack is clearly detected for the two coating thicknesses. Therefore, input power level maybe used as an additional parameter for enhancing crack detection under thick coatings.

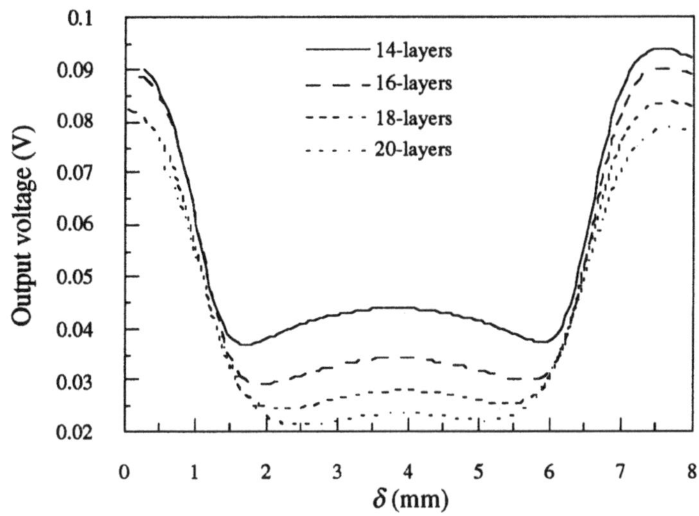

Fig. 4.43 Crack characteristic signals for a crack with width W = 0.51 mm and depth D = 2 mm at flush and when covered with four different layers of wrapping paper (coating thickness range of 0.56-0.80 mm) at 24 GHz (Zoughi *et al.*, 1995).

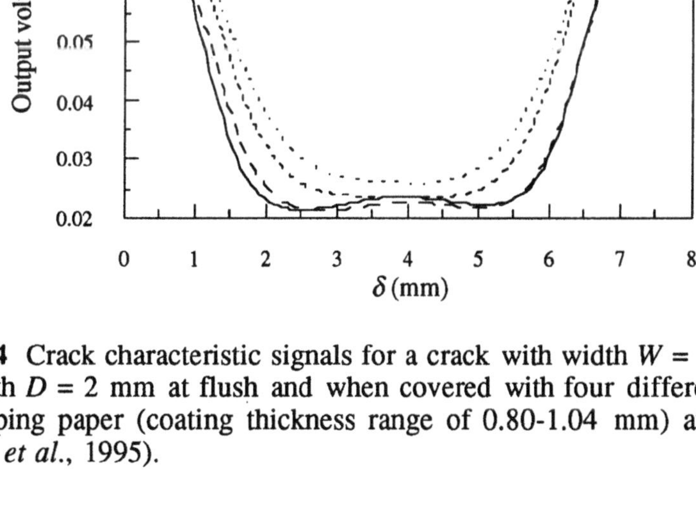

Fig. 4.44 Crack characteristic signals for a crack with width W = 0.51 mm and depth D = 2 mm at flush and when covered with four different layers of wrapping paper (coating thickness range of 0.80-1.04 mm) at 24 GHz (Zoughi *et al.*, 1995).

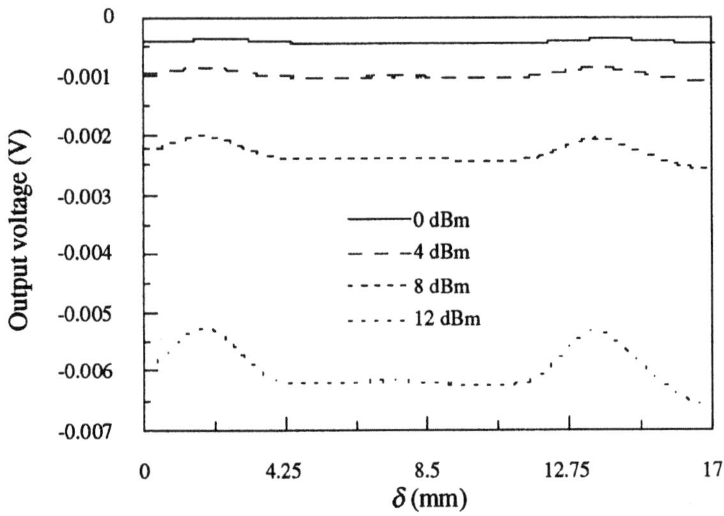

Fig. 4.45 Crack characteristic signals for a crack, in steel, with width W = 0.2 and depth D = 1.5 mm with eight layers of wrapping paper (coating thickness of 0.32 mm) at 17 GHz, for four different input power levels (Zoughi *et al.*, 1995).

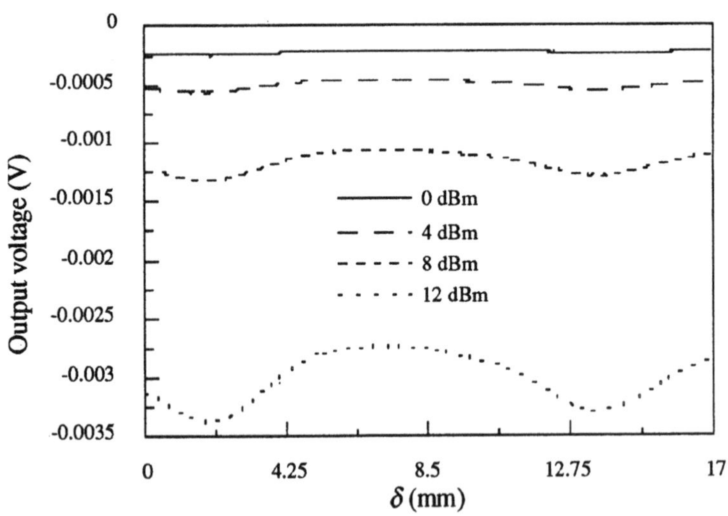

Fig. 4.46 Crack characteristic signals for a crack, in steel, with width W = 0.2 and depth D = 1.5 mm with thirty four layers of wrapping paper (coating thickness of 1.36 mm) at 17 GHz, for four different input power levels (Zoughi *et al.*, 1995).

This microwave approach is shown to be very powerful for detecting surface cracks under dielectric coatings such as paint, corrosion preventative coatings, etc. It was shown that cracks under coatings of greater than 1 mm are easily detected. Thin dielectric layers (≈<0.1 mm thick) may improve crack detection in two ways. The sharp transition becomes gradual (i.e. no need for high resolution scanners), and also in some cases there is an improvement in the dynamic range of the detected signal (if the location of the detector diode is fixed). Thus, in practice a thin layer of dielectric may be introduced, between the waveguide aperture and the specimen surface, to improve detection sensitivity particularly fixed diode positions. Furthermore, it was shown that the input power may be used as an additional optimizing parameter to probe surfaces with a few millimeters of dielectric coatings.

4.5.6 Remote crack detection (influence of liftoff)

Liftoff (standoff distance) between the waveguide aperture and the surface of a specimen is an important practical parameter. Remote crack detection requires incorporating a liftoff between the waveguide probe and the surface under examination. Liftoff could be viewed in three ways:

1. a finite thickness of air which may be present in most measurements or may be required (non-contact or remote measurements),
2. a finite thickness of air which may be used for crack detection optimization if its presence enhances crack detection,
3. a finite thickness of a dielectric with dielectric properties equal to those of free-space (i.e. covered crack case).

Therefore, it is important to study the effect of liftoff or airgap in our measurements.

Figures 4.47 and 4.48 show the crack characteristic signals recorded, at 24 GHz, for a crack with a width of 0.51 mm and a depth of 3 mm at flush (in-contact) and seven different liftoffs from 0.5 mm to 5 mm. There are two important features associated with these results. First, as the liftoff increases the distance between the two sharp transitions in the crack characteristic signal tends to increase (similar to covered cracks). This is as a result of the slot edges being "seen" by the microwave signal before they are over the waveguide aperture. Second, the level of the characteristic signals and their variations within the scanned distance seem to reduce markedly. This may be interpreted as the sensor (waveguide aperture) not being sensitive to the presence of the slot beyond a certain liftoff distance. However, this may not always be true as we will see shortly.

Another important feature illustrated in Fig. 4.47 is that if the liftoff changes between 1.5 mm and 2.5 mm (and perhaps even more) the detection sensitivity does not change very much. This means in practice an operator is not required to keep an accurate liftoff when detecting cracks in a remote fashion.

Since microwave signals are coherent in nature, one would expect that beyond a certain liftoff the signal may appear again. Figure 4.48 shows that once the liftoff is increased beyond 3 mm some impression of the characteristic signal begins to appear again, although not with the same distinct shape. This is an important observation particularly in light of the fact that the input signal power may be increased to detect a crack at farther distances from the aperture. Further investigations showed liftoffs in the order of 10 mm still render the crack detectable (Zoughi *et al.*, 1995).

Earlier, it was discussed in detail, the possibility of a crack not being detected depending on its depth. The crack would not be detected because the middle signal level in its characteristic signal and the signal when the crack is outside the waveguide aperture are similar. In such cases one may use liftoff as a parameter to make the crack detectable. Figure 4.49 shows the crack characteristic signal, recorded at 24 GHz, for a crack with a width of 0.51 mm and a depth of 6 mm at flush (in-contact) and at four liftoffs (0.5-2 mm). When considering the flush crack characteristic signal, one may conclude that the variation in its overall shape is small, and may not be sufficient for detection (this may not be true if the signal is normalized) (Zoughi, Ganchev and Huber, 1996). However, at 0.5 mm or 1 mm liftoffs the crack is clearly detectable (also for the other liftoffs). Figure 4.50 shows the same results except when normalized each with respect to its signal level when the crack is outside of the aperture. For liftoff of ≈1.5 mm the detection sensitivity is the highest. Beyond this liftoff the sensitivity decreases. Clearly, the results indicate that the introduction of liftoff causes the crack to become detectable.

The results of these investigations show the potential of remote (non-contact) surface crack detection with this microwave method. Normalized crack characteristic signals tend to identify the presence of a crack better than just the actual recorded crack characteristic signals. Furthermore, as discussed in the previous section, the level of input power may be increased to detect cracks at larger liftoffs. The results indicate that liftoffs in the range of a few millimeters to a few centimeters are indeed possible depending on the frequency of operation and the crack dimensions. It is also demonstrated that liftoff may make a crack detectable where it would not have been otherwise (depending on its depth). In addition, if the liftoff changes within a few millimeters (i.e. sensor is shaking so as to change the liftoff) the detection sensitivity may not suffer significantly. From a

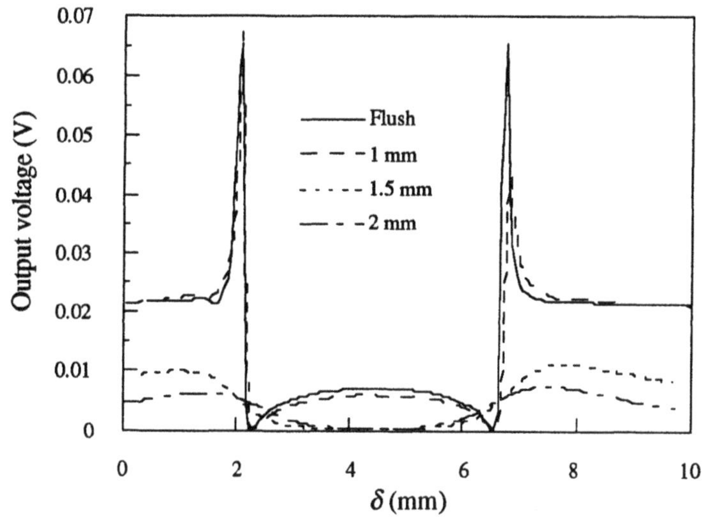

Fig. 4.47 Crack characteristic signals for a crack with width W = 0.51 mm and depth D = 3 mm at flush and three liftoffs (1-2 mm) at 24 GHz (Zoughi *et al.*, 1995).

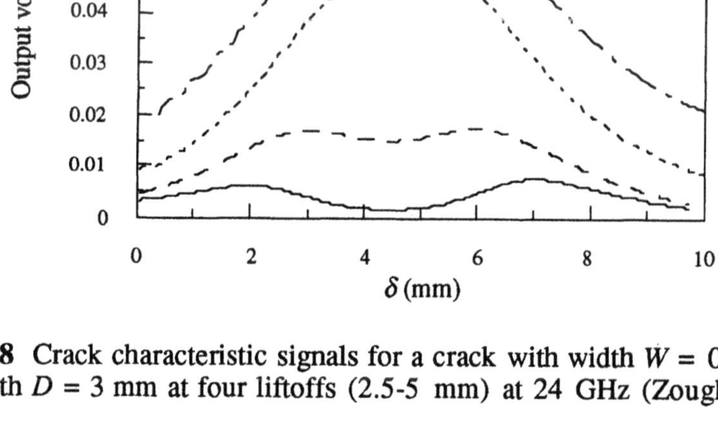

Fig. 4.48 Crack characteristic signals for a crack with width W = 0.51 mm and depth D = 3 mm at four liftoffs (2.5-5 mm) at 24 GHz (Zoughi *et al.*, 1995).

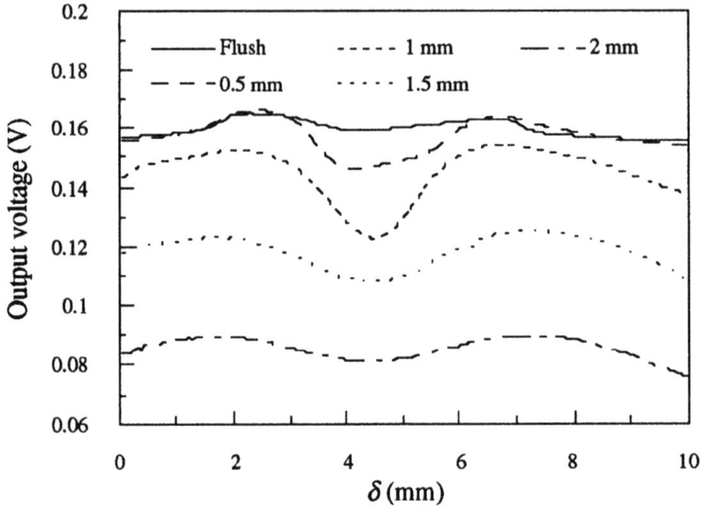

Fig. 4.49 Crack characteristic signals for a crack with width W = 0.51 mm and depth D = 6 mm at flush and four liftoffs (0.5-2 mm) at 24 GHz (Zoughi *et al.*, 1995).

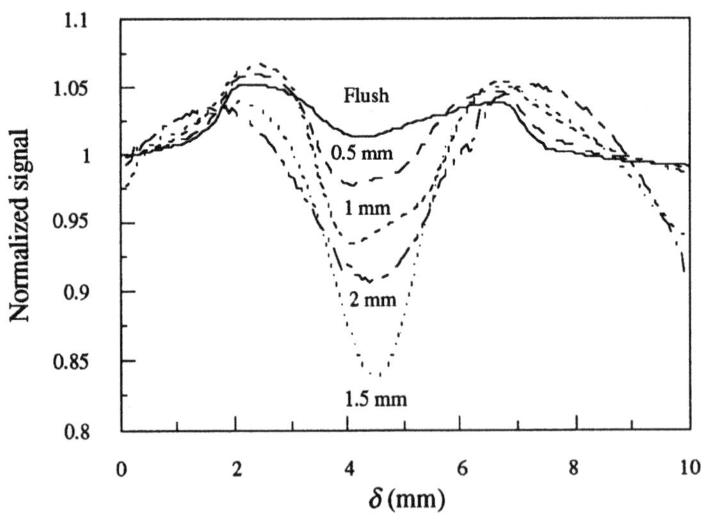

Fig. 4.50 Normalized crack characteristic signals for a crack with width W = 0.51 mm and depth D = 6 mm at flush and four liftoffs (0.5-2 mm) at 24 GHz (Zoughi *et al.*, 1995).

practical point of view, this is an important feature since it does not require a very rigid liftoff for remote crack detection. In a later investigation it was shown that larger liftoffs (in the order of a few millimeters) along with optimal incidence angles may also be incorporated into the measurement system (Huber *et al.*, 1997b). The incorporation of an incidence angle was shown to be very useful particularly when the metallic specimen under inspection is in motion. Finally, the introduction of a liftoff also causes the waveguide flange edges to interact with a crack as it is being scanned (Qaddoumi *et al.*, 1998b). This interaction is shown to effectively increase crack detection sensitivity.

4.6 CRACK SIZING[5]

Subsequent to detecting a surface crack using one of these techniques, sizing becomes the natural next step since sizing is very important for fracture analysis and repair considerations. The above models developed for surface crack detection and their associated numerical codes are used to establish a practical and useful crack sizing techniques (Yeh, 1995; Yeh and Zoughi, 1994a; Yeh and Zoughi, 1995).

The sizing method first estimates the crack width and depth from the dominant mode crack characteristic signal. The error associated with estimating the crack width and depth using this method is shown to be less than ±20%. For applications in which this is an excessive error a swept frequency technique may be used rendering much higher depth estimation accuracy. The swept frequency method relies on using the estimated crack depth using the crack characteristic signal for choosing the proper size of a waveguide. In this technique, by sweeping the operating frequency and by measuring the phase of the reflection coefficient at the waveguide aperture, the crack length and depth may be estimated accurately (Yeh, 1995; Yeh and Zoughi, 1995).

4.6.1 Crack width or opening estimation

The relative geometry of a waveguide with respect to a long crack was shown in Figs. 4.1 and 4.2 and with respect to a finite surface crack is shown in Fig. 4.51.

[5] Portions of this section are reprinted with permission from *Materials Evaluation*, vol. 53, no. 4, pp. 496-501, Copyright 1995 © The American Society for Nondestructive Testing, Inc.

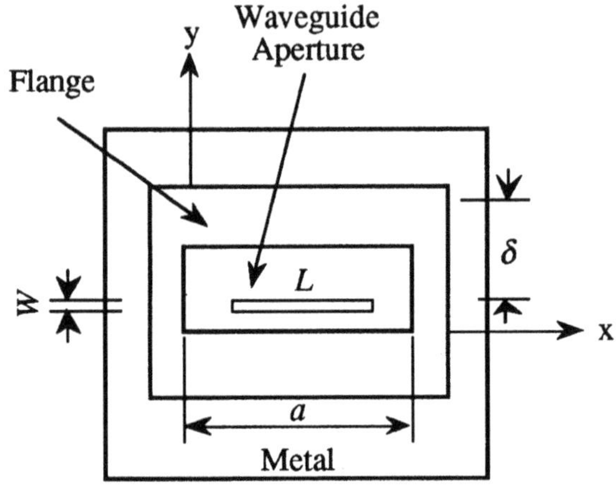

Fig. 4.51 Plan-view of a finite crack and a waveguide aperture (Yeh and Zoughi, 1995).

Crack width or opening is the dimension of the crack open to the waveguide aperture in the y-direction (see Figs. 4.2 and 4.51), crack depth is the dimension in the direction of wave propagation in the crack and the waveguide (in the z-direction, see Fig. 4.1) and the crack length is dimension in the x-direction (see Figs. 4.2 and 4.51). A long crack is one whose length is equal to or greater than the broad dimension of the waveguide, *a*, and a finite crack is one whose length is smaller than *a*. Two typical measured long crack characteristic signals are shown in Figs. 4.52 (referred to as a shallow crack) and 4.53 (referred to as a deep crack). These measurements were conducted at 24 GHz. The long and shallow crack in Fig. 4.52 has a width of 0.84 mm, a depth of D =1.53 mm and a length of 38 mm. The long and deep crack in Fig. 4.53 has a width of 0.94 mm, a depth of 10.08 mm and a length of 22.86 mm.

Using the experimental results for shallow cracks, the distance between the tips of the two outermost dips is approximately measured to be equal to the sum of the waveguide narrow dimension, b, and the crack width, W. Theoretically however, the distance between the points where the signal originally drops to its dip locations (Yeh and Zoughi, 1994a) is equal to (b + W). The difference between the experimental and the theoretical results is primarily due to the presence of a slight airgap or liftoff in between the waveguide aperture and the surface under examination. For narrower cracks, the liftoff effect is more significant because the perturbation caused by the crack becomes relatively less.

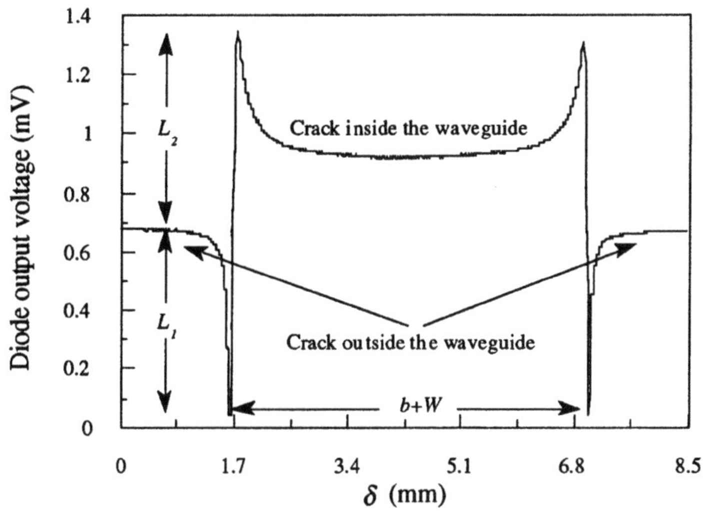

Fig. 4.52 Experimental crack characteristic signal for a relatively shallow crack (Yeh and Zoughi, 1995).

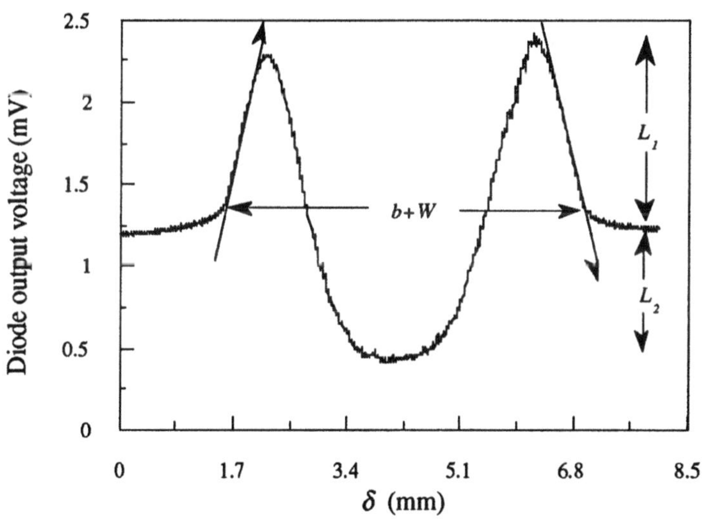

Fig. 4.53 Experimental crack characteristic signal for a relatively deep crack (Yeh and Zoughi, 1995).

The rounded corners associated with an actual crack (slot) also contribute to this difference. When the liftoff is small or negligible the crack width is under-estimated, while when the liftoff is large the crack width is over-estimated using this method. The significant peaks or dips in Fig. 4.52 occur when the crack is at the edges of the waveguide aperture, and the two roughly horizontal sections outside these peaks or dips represent the short circuit voltage value (i.e. when the crack is totally outside the waveguide aperture). As shown in Fig. 4.52, the distances from the short circuit value to the outermost dip (or peaks) and the following peak (or dip) are measured to be L_1 and L_2, respectively. Through conducting a large number of experiments a rule of thumb has been established which states that if the ratio of these two parameters, L_1/L_2, is not less than 0.76 or greater than 1.6, and if one takes the crack width as the distance between the tips of the two outermost dips minus the waveguide small dimension *b*, then the absolute error in estimating the width is less than 20%. For this ratio outside of this range, the liftoff or airgap between the probe and the test surface should be reduced and the experiment repeated until it in the above range (Yeh and Zoughi, 1995).

For deep cracks, the method for estimating crack widths is different than that described for shallow ones. As shown in Fig. 4.53, two tangential lines are drawn to overlap, as much as possible, the two edges of the two outermost dips or peaks. In practice, the distance between the two merging points is approximately the sum of the waveguide small dimension, *b*, and the crack width, *W*. Merging points are defined as those at which the characteristic signal, starting from the short circuit value, merges with the tangential lines. Similar to shallow cracks, this ratio for these cracks should be measured (L_1 and L_2 are distances as shown in Fig. 4.53). Extensive experiments have shown that the absolute error in width estimation is less than 20% if this ratio is less than 1.375.

Certainly, there may be some exceptions to these rules of thumb for experimentally estimating the width of shallow and deep cracks. However, crack width or opening may not be a particularly important parameter for fracture analysis or repair considerations. For these purposes, crack length and particularly depth information are sufficient for determining whether a specimen should be rejected (Yeh and Zoughi, 1995). A recent investigation has also shown similar relationships between crack characteristic signal features and crack width/opening (Qaddoumi *et al.*, 1998b)

4.6.2 Crack depth estimation

Theoretically, variations of $|E_y(z)|^2$ versus the scanning distance, δ, gives the crack characteristic signal (Yeh and Zoughi, 1994a). $|E_y(z)|^2$ is proportional to the voltage detected by a detector diode operating in its square law region (i.e. diode output voltage proportional to its input power) located at a certain distance away from the waveguide aperture. Theoretically, variations of $|E_y(z)|^2$ versus crack depth as shown in Fig. 4.54 can be obtained for the case when a crack is located at the center of the waveguide aperture. Three curves, for long cracks, with different widths of 0.5 mm, 0.25 mm, and 0.125 mm are plotted in this figure at 24 GHz. For a constant width one can see that $|E_y(z)|^2$ calculated at a distance of 9.45 cm away from the aperture first increases gradually with crack depth to a maximum value, then suddenly drops to zero around a specific depth (for that frequency), and after that increases gradually with crack depth again. From these three curves it is evident that if the crack width and length (long) are known, one can use the signal level in the middle part of the crack characteristic signal (i.e. crack in the middle of the waveguide) for depth estimation . Figure 4.54 also shows that the useable range (unambiguous depth determination) is up to one period of these curves for a given width. Theoretical results have shown that for a crack width of 40 microns if the error in width estimation is ±20% at 24 GHz the corresponding error in depth estimation using this approach is less than 18%. In practice, curves similar to those shown in Fig. 4.54 may be experimentally obtained for different crack widths and lengths before estimating crack depths since the output signal (measured voltage by the detector) is detector characteristic dependent. This dependence may be used to one's advantage to cause more separations between the curves for different widths. This method, however, is not extremely accurate because the signal level can be affected by detector characteristics and the airgap between the waveguide probe and the test surface. Therefore, this method is used for a rough estimation of crack depth only or in environments in which an error of less than 18% is acceptable. Higher accuracy in estimating crack depths can be obtained using a swept frequency approach.

4.6.3 Swept frequency technique

Since for fracture analysis the knowledge of crack width is not relatively important, only approaches for accurately estimating crack length and depth will be discussed in details. For simplicity, let us first assume that the crack length is equal to or longer than the broad dimension of the waveguide (i.e. a long crack). Finite cracks will be subsequently discussed.

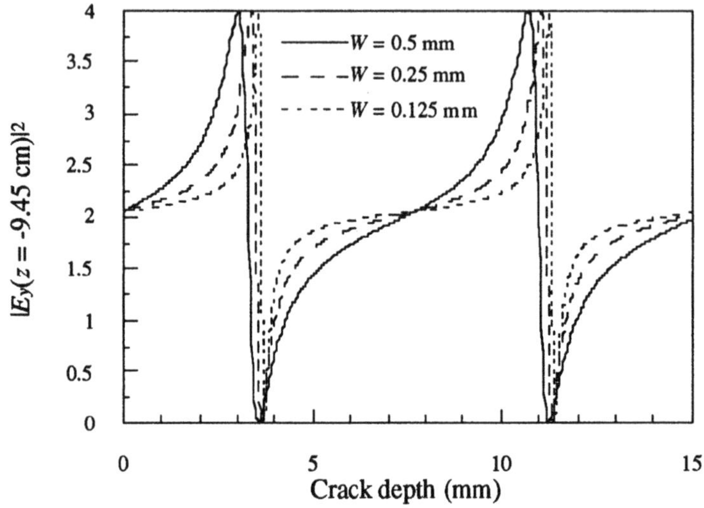

Fig. 4.54 $|E_y(z)|^2$ versus crack depth at 24 GHz for crack with widths W = 0.5 mm, W = 0.25 mm and W = 0.125 mm (Yeh and Zoughi, 1995).

As shown in Fig. 4.54, for cracks with a length equal to the broad dimension of the waveguide, *a*, the first abrupt transition occurs when the crack depth is approximately equal to one quarter of the guide wavelength, and then repeats every half guide wavelength thereafter. This is due to the phase variation of the reflection coefficient, at the waveguide aperture. The first abrupt transition is frequency dependent since it always occurs at a depth of about one quarter of the guide wavelength. If similar curves are plotted as a function of frequency, the transitions will occur at different depths for different frequencies. This means that one can sweep the operating frequency to obtain the transition frequency for a crack, and then subsequently measure its depth. The detected characteristic signal level depends on the input power-output voltage characteristics of the crystal detector and its location along the waveguide (Yeh and Zoughi, 1994a). Therefore, theoretical curves can not be used for crack sizing directly. However, if one measures the phase of the reflection coefficient the dependence on the detector characteristics is eliminated, and a direct comparison between the experimental and the theoretical results can be made (Bakhtiari et al., 1994). Figure 4.55 shows the phase of the reflection coefficient at the waveguide aperture as a function of crack depth with a crack with a width of 20 micrometers while the crack is at the center of the waveguide aperture for three different frequencies of 100 GHz, 38 GHz, and 24 GHz. From this figure it is clear that for cracks of the same width but different depths the abrupt transition of the phase of the reflection coefficient occurs at different frequencies.

Suppose the crack has a width equal to the small dimension of the waveguide aperture, *b* (unrealistic, but used here only for discussion purposes), the phase of the reflection coefficient is then a linear function of the crack depth. In this case, when the crack depth is exactly equal to one quarter of the guide wavelength the phase of the reflection coefficient is 0° which is right at the center of the phase transition in Fig. 4.55. However, for practical cases where the crack (slot) width is narrower than *b*, the change in the phase of the reflection coefficient becomes less abrupt as a function of crack depth. Since there is an induced current flowing along the crack surface, a mutual inductance exists between the two faces of the crack. The interaction between these currents causes a phase lag (maximum 90°) in addition to the phase lag which is a linear function of the crack depth (i.e. $2D$). For narrower cracks this inductance is small until the crack depth reaches one quarter of the guide wavelength so that the currents on the two crack faces are in phase. In other words, for a narrow crack, when the crack depth reaches a certain value the phase of the reflection coefficient changes more abruptly than that for a wider crack, and this phase transition occurs at a larger depth. Figure 4.56 shows three such transitions at 24 GHz for three cracks with widths of 20 micrometers, 50 micrometers and 200 micrometers, respectively. It is evident that the transition is steeper for narrower cracks. Since the phase of the reflection coefficient for a narrower crack has a steeper transition characteristics and shifts to the right, it intersects with that for a wider crack at the lower part of the curve (e.g. around -115° in Fig. 4.56). From Fig. 4.56 one can see that the phase is not so sensitive to the variation of crack width. In other words, for a negative phase angle, the depth corresponding to a width of 20 micrometers is not much different (in percentage) from that for a width of 50 micrometers or 200 micrometers. This is particularly true if the operating frequency is tuned such that a phase value of approximately close to -115° is detected. The reason why one can tune the frequency to achieve the desired phase angle for the reflection coefficient is clear from Fig. 4.55.

If the top or the bottom portions of these curves are used for depth estimation the resolution is about 0.75 mm or 0.5 mm per 30° (not linear), respectively. However, for the middle part of these curves the resolution is about 0.1 mm per 300° (nonlinear) which is much higher. Hence, if high accuracy is required one may sweep the testing frequency until an abrupt transition in phase is observed. Then measure the phase of the reflection coefficient at a frequency around the transition and estimate the crack depth using some calibration curves similar to those shown in Fig. 4.56. As shown in Fig. 4.56, for different crack widths, the slopes of these curves are different and thus their corresponding crack depth estimation accuracy.

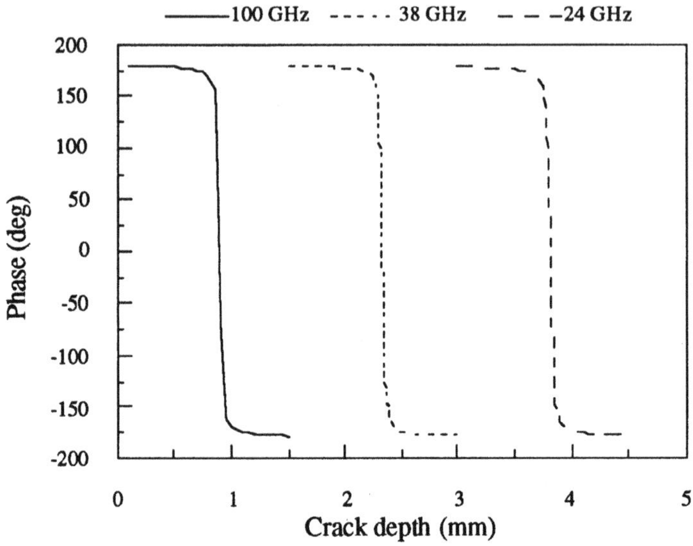

Fig. 4.55 Phase of reflection coefficient versus crack depth for cracks with equal widths $W = 0.2$ mm at frequencies of 100 GHz, 24 GHz and 38 GHz (Yeh and Zoughi, 1995).

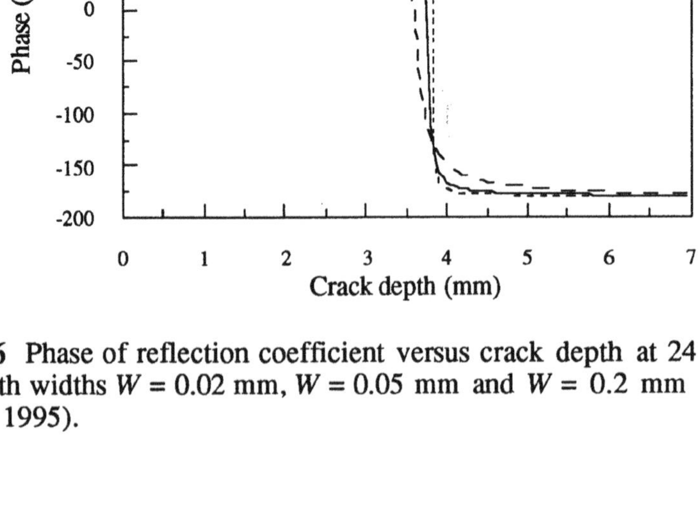

Fig. 4.56 Phase of reflection coefficient versus crack depth at 24 GHz for crack with widths $W = 0.02$ mm, $W = 0.05$ mm and $W = 0.2$ mm (Yeh and Zoughi, 1995).

If the swept frequency technique is needed to satisfy the accuracy requirements, five different waveguide sizes, W-band (75-110 GHz), V-band (50-75 GHz), U-band (40-60 GHz), Ka-band (26.5-40 GHz), and K-band (18-26.5 GHz), are required to cover the depth range from 0 to 4 mm for narrow cracks. The wider a crack gets, the fewer number of waveguides are required. If a crack has been detected and has a depth shallower than the range covered by the waveguide used in detection, one may fill the crack with a known dielectric such as dye penetrant to make the crack effectively deeper so that the transition in phase may occur at a lower frequency. Hence, if a crack is detected by the dye penetrant testing method, such a microwave technique might be used to estimate its depth making these two techniques complimentary (Yeh, 1995; Yeh and Zoughi, 1995).

4.6.4 Influence of crack length on the phase of reflection coefficient

In the above discussions, the crack length was assumed to be equal to or longer than the broad dimension of the waveguide. The effect of crack length (e.g. finite cracks) on the phase of the reflection coefficient is investigated next. Having positioned a finite crack in the middle of a waveguide aperture, Fig. 4.57 shows the phase of the reflection coefficient as a function of crack depth for cracks with a width of 200 micrometers but for three different lengths of 7 mm, 6 mm, and 5 mm at 38 GHz. These three curves are similar in shape to those for long cracks as shown in Fig. 4.56. Both experimental and numerical results have shown that when the crack length is shorter than *a*, the curve shifts to the right because shorter cracks are less effective in shifting the phase of the reflection coefficient, and a greater crack depth is needed to compensate for this. Similarly, if the crack width and depth are fixed, changes in crack length also shift the phase of the reflection coefficient in the same way when the operating frequency is swept. The three curves shown in Fig. 4.58 are similar to those in Fig. 4.56, but for three finite cracks. These cracks have equal lengths of 6 mm, but different widths of 200±20% micrometers, respectively at 38 GHz. From this figure one can see that the observations made for long cracks (referring to Fig. 4.56) are also true for finite cracks. In other words, the influence of the error in crack width estimation on depth estimation can also be minimized for finite cracks by properly choosing the operating frequency. If both the crack width and length are known, then curves similar to those shown in Fig. 4.58 can be used for crack depth estimation (Yeh, 1995, Yeh and Zoughi, 1995).

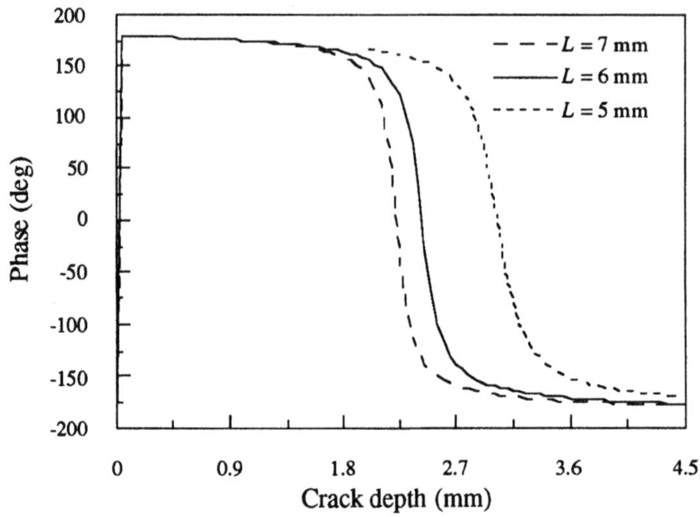

Fig. 4.57 Calculated curves of the phase of reflection coefficient versus crack depth for three cracks with widths W = 0.2 mm and depth D = 2.5 mm, but with three different lengths of 5 mm, 6 mm and 7 mm (Yeh and Zoughi, 1995).

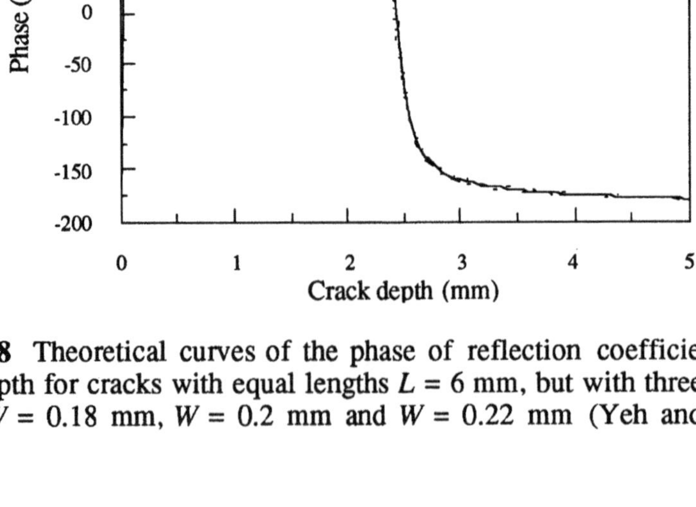

Fig. 4.58 Theoretical curves of the phase of reflection coefficient versus crack depth for cracks with equal lengths L = 6 mm, but with three different widths W = 0.18 mm, W = 0.2 mm and W = 0.22 mm (Yeh and Zoughi, 1995).

4.6.5 Crack length estimation

For a long crack the phase of the reflection coefficient remains constant as the waveguide probe moves along the crack length (i.e. in the x-direction). However, when either end of the crack moves into the waveguide aperture such that the crack becomes finite, the phase of the reflection coefficient changes abruptly. This is as a result of the fact that the dominant waveguide mode, TE_{10}, can propagate into the crack since for this mode to propagate inside a waveguide, its broad dimension must be equal to a (which is the case for a long crack). While, for finite length crack cases the dominant waveguide mode (not the same as the dominant mode of the crack) can no longer propagate into the crack (i.e. crack length less than a) and the wave composition (wave modes and their associated magnitudes and phases) changes abruptly. This change causes the significant shift in the phase of the reflection coefficient (see Fig. 9 of Yeh and Zoughi, 1995). This phenomenon may be used to accurately determine the ends of a finite crack and hence its length, as will be seen later.

4.6.6 Experimental results

Several experiments were conducted to illustrate the applicability of these crack sizing techniques particularly the swept frequency depth measurement (Yeh, 1995; Yeh and Zoughi, 1995). To conduct the swept frequency measurements, a microwave bridge consisting of a dual-arm microwave reflectometer was constructed as shown in Fig. 4.59. The sweep oscillator generates the swept frequency microwave signal which is then fed into a microwave tee. The tee splits the input signal into two equal (phase and magnitude) signals each entering one of the reflectometer arms. In the upper arm the signal is fed into an open-ended (probing) waveguide in contact with the specimen under test. The reflected signal is subsequently coupled into the attached directional coupler, and is finally fed to the test port of a microwave network analyzer. The other half of the input signal enters the lower arm, and is reflected off of a sliding short circuit. The reflected signal in this arm is then fed into the reference port of the network analyzer for comparison (phase and magnitude) with the test signal. The sliding short in the reference arm provides compensation for the extra length of the probing waveguide in the test arm and any other unwanted reflections over the entire frequency range. Once this set-up is calibrated with respect to a short circuit at the test port, any detected change in the measured parameters will be due to the crack within the waveguide aperture. The information (phase and magnitude) measured by the network analyzer can be monitored by a computer through an A/D converter board which makes real-time processing possible. An HP8410 microwave network analyzer (operating frequency range of 8-12.4 GHz) was used in these experiments. Due to the relatively low frequency range of the network analyzer the depth of cracks that can be used to test the feasibility of the swept frequency method is around 10 mm. Clearly, such a crack may not be a realistic one, however using an HP8510 or HP8720 series network analyzer will allow inspecting cracks with depths of less than one millimeter. The goal of these experiments were to show the feasibility of the technique. It must be noted that in a practical setting one would build a specialized microwave testing system/hardware as opposed to using a network analyzer which is relatively expensive.

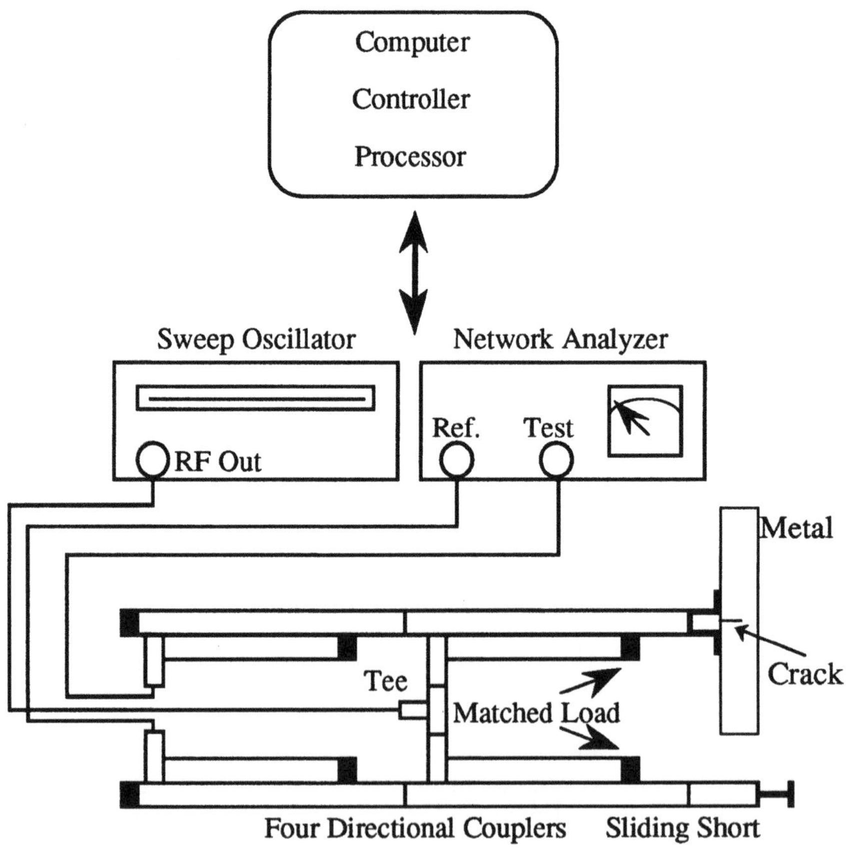

Fig. 4.59 Waveguide bridge for swept frequency measurement of the phase of reflection coefficient (Yeh and Zoughi, 1995).

To experimentally examine the swept frequency approach two long slots were milled on an aluminum plate with widths of 0.936 mm, but with depths of 10.08 mm and 9.82 mm, respectively. To measure the width of these slots, their characteristic signals were used at 24 GHz, as explained earlier. The characteristics signal for the 10.08 mm deep crack is the one shown in Fig. 4.53. Applying the crack width estimation technique outlined for deep cracks, the width of this slot was measured to be 1.03 mm. The error associated with the width estimation is +10%. By the shape of the curve shown in Fig. 4.53, it is evident that the crack is a deep one at 24 GHz which means lower frequencies should be used for sizing its depth. Hence, the microwave bridge described above was used to determine the crack/slot depth. Figure 4.60 shows the results of the swept frequency measurement and the results of the theoretical code (Yeh and Zoughi, 1994a) for these two cracks/slots. At lower frequencies, the difference between the experimental and theoretical results may be partly due to the calibration error, while at the higher frequencies these curves overlap quite well. This confirms the statement made earlier that theoretical curves of the phase of the reflection coefficient versus frequency can be used directly for crack sizing. The frequencies around the lower turning points, where the curves change their slopes, are those at which crack depth estimation should be performed. For example, for the crack with a depth of 10.08 mm, frequencies of around 9.5 GHz are adequate, while for the depth of 9.82 mm, a frequency of 9.8 GHz may be better. For the crack with the depth of 10.08 mm at 9.5 GHz the phase of the reflection coefficient is -55°. Since the error in width estimation is not known in practice (however, in this example the error was measured to be +10%) let's assume a maximum error of ±20%, and investigate the effect of this error on depth estimation. Figure 4.61 shows three theoretical curves of the phase of the reflection coefficient versus crack depth at 9.5 GHz for crack widths of 0.936 mm, 0.936+20% mm and 0.936-20% mm, respectively. Monitoring the phase angle at 9.5 GHz (-55°) the three corresponding estimated depths are: 10.079 mm (negligible error) for an estimated width of 1.123 mm; 10.110 mm (an error of +0.3%) for an estimated width of 0.936 mm; and 10.171 mm (an error of +0.9%) for an estimated width of 0.749 mm. This approach clearly shows the effectiveness of the swept microwave frequency measurements for accurate crack/slot depth determination.

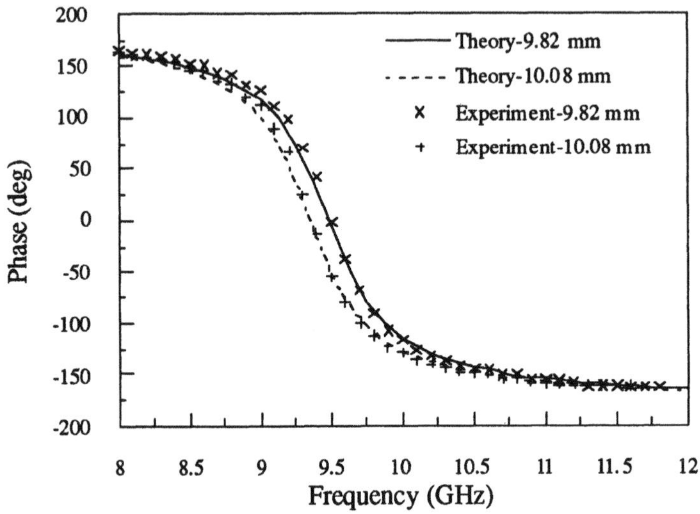

Fig. 4.60 Comparison of the theoretical and experimental results of the phase of reflection coefficient versus frequency for cracks with widths W = 0.939 mm and depths D = 9.82 mm and D = 10.08 mm, respectively (Yeh and Zoughi, 1995).

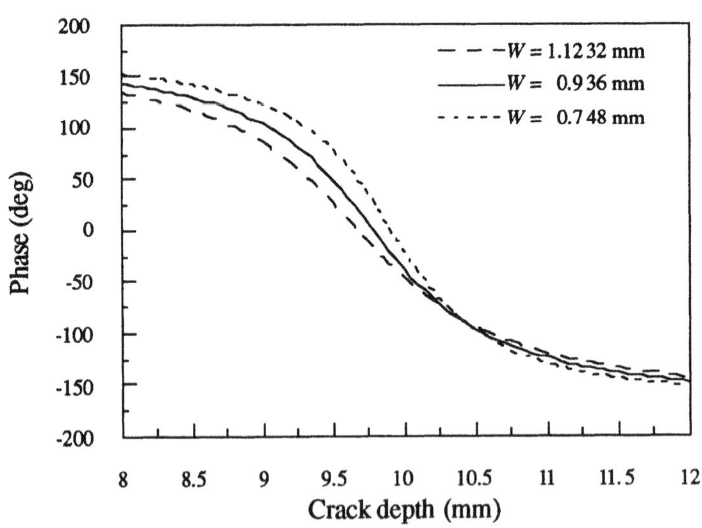

Fig. 4.61 Phase of reflection coefficient versus crack depth at 9.5 GHz for crack with widths W = 0.936 mm, W = 0.748 mm and W = 1.123 mm, respectively (Yeh and Zoughi, 1995).

4.7 TIP LOCATION DETERMINATION[6]

For repair purposes it is often necessary to know the exact location of the tip of a propagating crack. One practice for preventing crack propagation is to drill a hole at the crack tip. The crack characteristic signal is unique for a given crack dimensions, operating frequency and waveguide dimensions. Several typical normalized crack characteristic signals for various crack widths and depths were shown in Figs. 4.22 and 4.23. The difference between the signal levels in the middle of the crack characteristic signals can be used for depth determination (Yeh and Zoughi, 1995). To better understand crack tip location identification it is important to note that the signal level when the crack is inside the waveguide aperture is a function of crack depth and width (in that order of importance). Furthermore, the distance between the two sharp transitions is a function of the narrow dimension of the waveguide and the width of the crack (although depth influences this distance as well but less significantly) (Zoughi *et al.*, 1995). Also, when the crack (particularly a shallow one) is inside the waveguide aperture and away from the edges of the waveguide, the signal level remains fairly constant (except for relatively deep cracks, as shown in Section 4.5).

Assuming a crack has been found, there are two methods to determine its crack tip location. The first method involves scanning a crack in two directions (two-dimensional crack characteristic signal) and producing its image. The second method requires scanning the crack only in one direction (**crack tip characteristic signal**). The latter approach is more useful for practical applications since only scanning in one direction is required to identify crack tip location. However, the former approach is useful for understanding the process of tip location identification as well as obtaining general information abut the crack (Zoughi, *et al.*, 1995; Ganchev *et al.*, 1996).

Although a 2-D crack characteristic signal may be interesting since it shows the impression of a crack as it appears on a metal surface, practically its production is time consuming and somewhat cumbersome. Consequently, an alternative simple and fast approach for identifying crack tips was developed. Once a crack has been detected, the waveguide aperture may be placed on it as shown in Fig. 62a. For this case it is sufficient that the crack be aligned somewhere in the middle of the narrow dimension of the waveguide (as will be discussed later). Now if the waveguide only moves along the x-direction (as depicted sequentially in Figs. 62b-c), first the crack is totally inside the aperture, then its tip enters the aperture where a significant change in the detected signal should occur then it continues to

[6] Portions of this section are reprinted with permission from *Materials Evaluation*, vol. 54, no. 5, pp. 598-603, Copyright 1996 © The American Society for Nondestructive Testing, Inc.

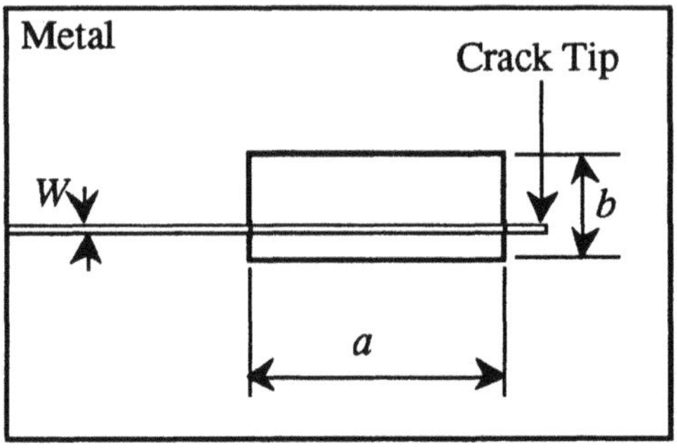

Fig. 4.62a Relative geometry of a waveguide aperture and a crack when the crack is fully within the aperture (Ganchev *et al.*, 1996).

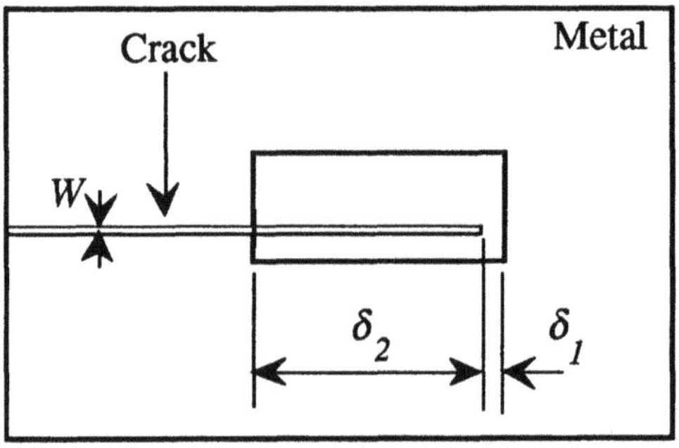

Fig. 4.62b Relative geometry of a waveguide aperture and a crack when the crack tip having just entered inside the aperture (Ganchev *et al.*, 1996).

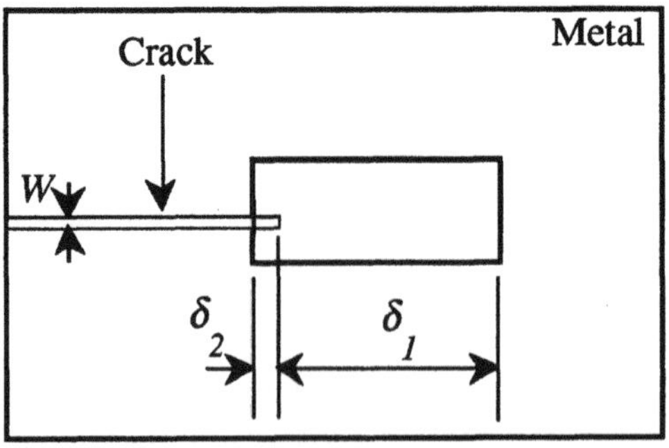

Fig. 4.62c Relative geometry of a waveguide aperture and a crack when the crack tip is about to leave the aperture (Ganchev *et al.*, 1996).

move further inside the aperture where the change in the detected signal becomes gradual, and finally the crack tip leaves the aperture. When the crack is fully within the waveguide aperture (Fig. 62a), the dominant TE_{10} mode in the waveguide and in the crack are the same. This is due to the fact that the length of the crack inside the aperture is equal to the broad dimension of the waveguide (Yeh, 1994). When the crack tip enters the waveguide aperture (Fig. 62b), the dominant mode in the crack is no longer the same as that in the waveguide. This has a relatively significant (abrupt) influence on the crack characteristic signal. As the crack continues to be partially inside the waveguide aperture the change in the crack characteristic signal becomes gradual until the crack is totally outside the aperture, in which case a constant voltage is detected (short circuit case). Therefore, by recording the detected voltage as a function of the scanning distance along x-direction the crack tip characteristic signal is obtained. The results are better illustrated when the percent absolute difference between the measured voltage, at any point in the scan, and when the crack is totally outside the aperture (short circuit case) is plotted as a function of the scanning distance. In this section scanning distance refers to movement of the waveguide aperture along the x-direction.

Figure 4.63 shows the crack tip characteristic signal, obtained in this manner, for a slot with a width of 0.51 mm and a depth of 1.5 mm at 24 GHz. The three distinct regions in this figure show the location of the slot, with respect to the waveguide aperture, along the x-direction. The two vertical lines indicate the relative location of the slot tip once at one edge of the waveguide and once at the other, respectively. Thus, the distance between these two lines is equal to the broad dimension of the waveguide aperture, a. The results show that when the crack is fully within the waveguide aperture the detected-voltage-change is relatively constant, as expected. However, as soon as the crack tip enters the aperture (indicated as "crack tip" in the figure) this detected-voltage-change abruptly decreases and continues to decrease while the crack tip moves along inside the aperture. When the crack comes to the vicinity of the other side of the waveguide (≈3 mm away, which is about a third of the broad dimension of the waveguide) there is hardly any variation in the detected-voltage-change, also as expected. The theoretical model, describing the interaction of a crack with an open-ended rectangular waveguide was used to theoretically obtain the results of this experiment (also depicted in Fig. 4.63) (Huber *et al.*, 1997a). Clearly, there is excellent agreement between the theoretical and the experimental results. From this experiment and many similar ones (for different crack dimensions) it is concluded that the crack tip position may be identified accurately to within 0.25 mm of its actual location (0.25 mm is also the scan step size used, along x-direction, to obtain the results shown in Fig. 4.63). A finer scan step size may result in a more accurate crack tip location identification.

The above experiment was conducted while the crack was positioned in the middle of the narrow dimension of the waveguide. Figures 4.22 and 4.23 shows that as long as the crack (a relatively shallow one) is within the narrow dimension of the waveguide (excluding right at the edges) the detected voltage remains fairly constant. Thus, the placement of a crack exactly in the middle of the narrow dimension of the waveguide is not necessary for identifying crack tip position when using this microwave methodology. To illustrate this fact, the crack tip characteristic signal for a crack with a width of 0.58 mm and a depth of 2 mm was obtained at 24 GHz while placing the crack at four different locations around the middle of the narrow dimension of the waveguide (0.5 mm apart), as shown in Fig. 4.64. The results are indeed as expected (two of them are actually on top of each other). This has important practical ramifications since this microwave approach does not require accurate placement of the waveguide over the crack along the y-direction.

The influence of different crack dimensions, on the crack tip characteristic signals, is investigated next. Figure 4.65 shows the crack tip characteristic signal, recorded at 24 GHz, for three cracks with equal widths of 0.51 mm and depths of 1.5 mm, 2.5 mm and 3 mm, respectively. As expected, crack depth variations cause a change in the level of the detected signal when the crack is totally within the waveguide. Likewise, for all three cases shown in Figure 4.65 the short circuit level is the same (i.e. detected-voltage-change for scanning distances greater than 19.5 mm), but the signal level for different cracks, when fully within the waveguide aperture is different (i.e. detected-voltage-change for scanning distances between 7 mm to 8.5 mm), as expected (Yeh, 1994; Zoughi *et al.*, 1995). The accuracy of the crack tip location identification therefore will depend on the difference between the short circuit signal level (when the crack is outside the waveguide aperture) and the signal level when the crack is fully inside the aperture. For example for D = 1.5 mm one may argue that the tip is located within ≈1.25 mm of its actual location, whereas for D = 2.5 mm the accuracy may be within 0.25 mm. However, if the crack tip characteristic signal for D = 1.5 mm is plotted by itself (i.e. changing the scale of the graph) then the location of the abrupt change (i.e. tip location) is within ≈0.25 mm to the actual crack tip location. In any case, this is not necessarily a drawback since once the crack is detected, the operating frequency or the detector diode position may be adjusted to maximize the signal difference between the two levels (Zoughi, et al., 1995, Zoughi, Ganchev and Huber, 1996). Figure 4.66 shows crack tip characteristic signals for three cracks, recorded at 24 GHz, with equal depths of 2 mm and widths of 0.41 mm, 0.51 mm and 0.58 mm, respectively. For these

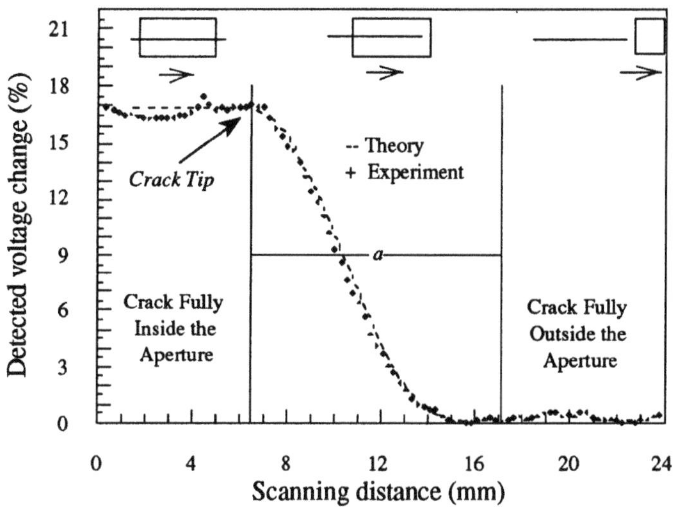

Fig. 4.63 Calculated and measured crack tip characteristic signals for a long crack with width W = 0.51 mm and depth D = 1.5 mm at 24 GHz (Ganchev *et al.*, 1996).

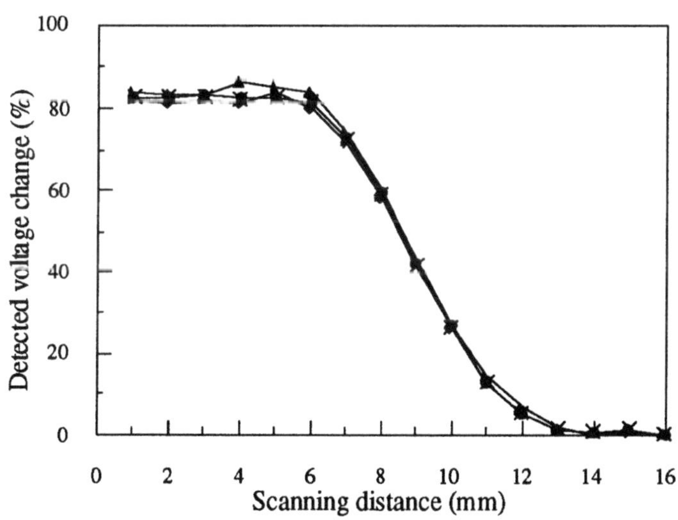

Fig. 4.64 Crack tip characteristic signals for a crack with width W = 0.85 mm and depth D = 2 mm at 24 GHz measured at four positions (each 0.5 mm apart) around the middle of the narrow dimension of the waveguide aperture, b (Ganchev *et al.*, 1996).

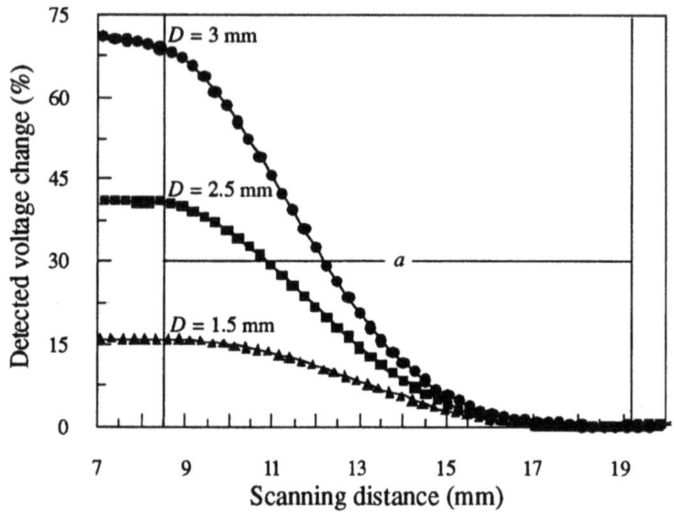

Fig. 4.65 Crack tip characteristic signals for three cracks with equal widths $W = 0.51$ mm and three different depths at 24 GHz (Ganchev *et al.*, 1996).

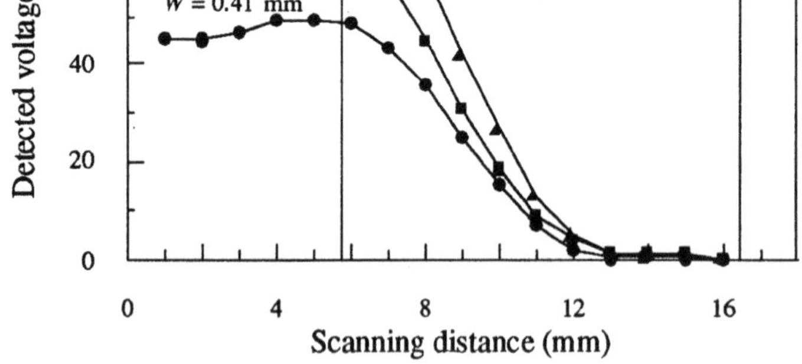

Fig. 4.66 Crack tip characteristic signals for three cracks with equal widths $D = 2$ mm and three different widths at 24 GHz (Ganchev *et al.*, 1996).

cracks the tip locations are identified to within ≈1 mm (which is the scan step size in these measurement). As in the previous case the larger the difference is between the signal levels when the crack is outside and inside the aperture, the more accurate the crack tip may be located.

Next, the crack tip characteristic signals for an empty and a filled (with beeswax) crack are compared. We have shown that when a crack is filled its electrical depth changes (Yeh and Zoughi, 1995; Zoughi, *et al.*, 1995). Thus, the signal level when the crack is fully inside the aperture will change. This means that the crack tip characteristic signal for a filled crack (compared to an empty crack) should be similar to a crack whose depth is larger. Figure 4.67 shows the results of an experiment in which a crack with a width of 0.3 mm and a depth of 2 mm was filled with beeswax and the crack tip characteristic signals for this crack when empty and when filled were recorded at 24 GHz. The results follow the expectation of the filled crack now having a larger depth (D).

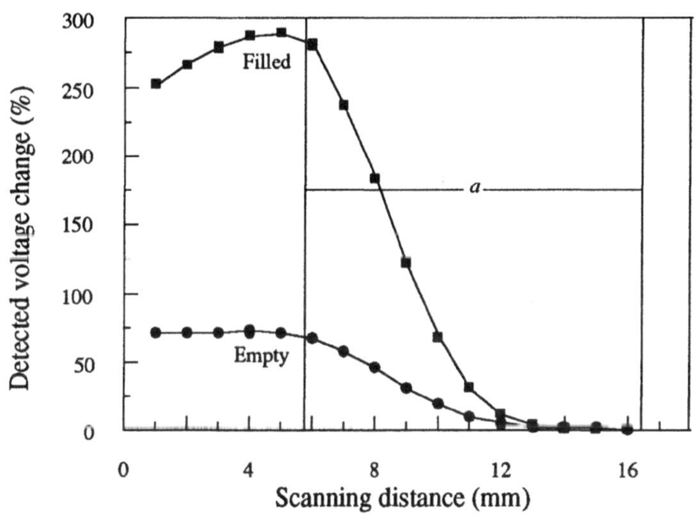

Fig. 4.67 Crack tip characteristic signals for an empty and filled (with beeswax) crack with a width $W = 0.3$ mm and depth $D = 2$ mm at 24 GHz (Ganchev *et al.*, 1996).

The crack tip characteristic signal for covered cracks was also investigated, as well as the accuracy of determining the crack tip location. Crack tip characteristic signals for a crack with a width of 0.51 mm and a depth of 1.5 mm, recorded at 24 GHz, when the crack was covered with 2 sheets (0.08 mm), 6 sheets (0.24 mm), 12 sheets (0.48 mm), 16 sheets (0.64 mm) and 20 sheets (0.8 mm) of wrapping paper (simulating various paint thicknesses since wrapping paper and common paint have similar dielectric properties) are shown in Fig. 4.68. The relatively abrupt transition around 9 mm still exists for all coating thicknesses. For these covered crack cases, the accuracy by which the tip of a crack may be located is estimated to be within ≈2 mm of its actual position. As explained earlier, the accuracy of determining covered crack tip locations may also be improved by optimizing the operating frequency and the detector diode position. However, for all cases discussed in this paper, the accuracy by which the tip position of a crack may be located is very good.

It must be noted that the results presented here are raw data without the application of any post processing. Simple processing routines such as forward difference method for gradient approximation (i.e. calculating forward slope for every data point) may be another way to analyze the data. Figure 4.69 shows the forward slope calculated for each point of the data shown in Fig. 4.65. It may be suggested that, using this approach, the crack tip is identified when the slope begins to change substantially. "Substantially" will then depend on the particular application, level of accuracy required, and whether the crack is covered or filled.

In summary, a one dimensional scan, for obtaining the crack tip characteristic signal, is shown to be a powerful approach for quick and accurate crack tip position identification. The characteristics of the detected signal when the crack is fully within the waveguide aperture and when its tip enters the aperture is used to produce the crack tip characteristic signal. The accuracy of the crack tip location for empty and filled cracks is estimated to be within ≈ 0.25 mm of the actual tip location. If necessary, the accuracy may be improved by changing the operating frequency or the detector diode position. In the case of covered cracks this accuracy is less, depending on the thickness of the coating, and was shown to be within ≈2 mm of its actual location. Experimental and theoretical results were also compared indicating excellent agreement between the two. Thus, the theoretical code may be used as an optimizing tool for enhancing the accuracy of crack tip location identification.

4.8 DETECTION OF STRESS-INDUCED FATIGUE CRACKS

To demonstrate the practical utility of this microwave crack detection approach an A-36 steel specimen with a thickness of 12.7 mm and with the dimensions shown in Fig. 4.70 was used in which a small hole was drilled through the specimen and a sawcut starter notch was introduced on both sides of the hole as shown in Fig. 4.71 (Qaddoumi *et al.*, 1998a, Qaddoumi *et al.*, 1998b). The specimen was then fatigued at 8 Hz, using a closed loop servo-hydraulic fatigue machine, at a maximum stress of 103 Mpa (15 ksi) and a stress ratio of 0.05. Consequently, a tight fatigue crack was generated

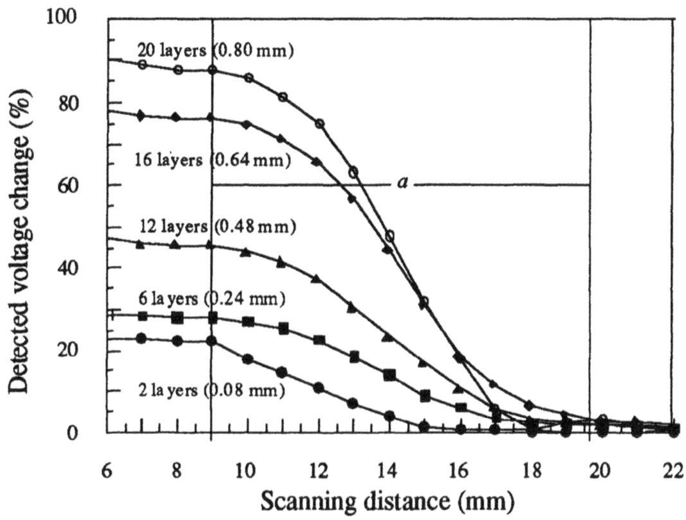

Fig. 4.68 Crack tip characteristic signals for a crack with width $W = 0.51$ mm and depth $D = 1.5$ mm at 24 GHz while covered with up to 20 layers of wrapping paper (Ganchev *et al.*, 1996).

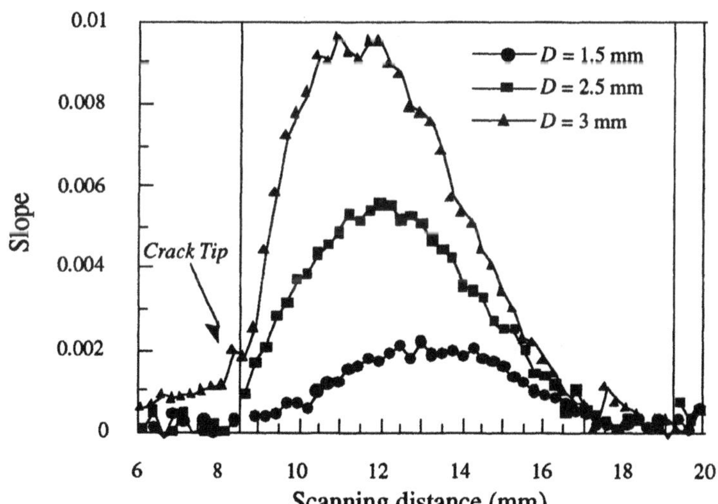

Fig. 4.69 Forward point-to-point slope calculated for the crack shown in Fig. 4.65 (Ganchev *et al.*, 1996).

at the end of the notch whose width varied as a function of the distance away from the notch tip, as shown in Fig. 4.72. The dominant mode measurements were made at a point half way between the tip of the machined starter notch and the final crack tip. As the crack tip advances during fatigue, the surface of the specimen (at the crack tip) is in a state of plane stress, which influences the size of the plastic zone. The crack front typically tunnels along its path, with the specimen surface crack trailing the inner portion of the crack front. The uncracked ligament between the inner portion of the crack front and the specimen surface yields as the crack advances, creating a slight indentation along the crack length. This indentation depth varies with the size of the plastic zone. When the plastic zone is large, the indentation can be large, (analogous to necking in a tensile specimen). But when the plastic zone is small, as in fatigue of bridge structures, the indentation can be quite small. The indentation on this A-36 steel sample was very small, and could not accurately be measured without destroying the sample. An estimate of the indentation, using a replication technique, showed the depth to be less than 0.040 mm. Figure 4.73 shows the crack opening or width as a function of loading, measured halfway between the starter notch tip and the fatigue crack tip.

Using a custom made dominant mode probe operating at a frequency of 24 GHz, this crack was scanned while under various loading (Qaddoumi *et al.*, 1998b). Figure 4.74 shows the results of these experiments. The results clearly indicate that the crack is detected even at low loading values, and the dynamic range of the detected signal increases as the loading increases. The increase in loading causes the crack opening to increase (and to some extent its depth). Thus, the respective crack characteristics signals become more pronounced. The fact that the short circuit levels for each loading is different is due to the different liftoffs that are created in between the probe and the steel specimen as the loading increases. The quantitative analysis of detection sensitivity as a function of loading or crack opening is an ongoing effort at the time of writing of this text.

4.9 SUMMARY

In this chapter the many feature and attributes of a novel microwave technique for detecting and evaluating the properties of many slots and cracks including stress-induced fatigue surface cracks in metals was presented. Microwave techniques, using open-ended rectangular waveguide aperture probes, offer certain advantages when detecting hair-line stress or fatigue cracks including the following:

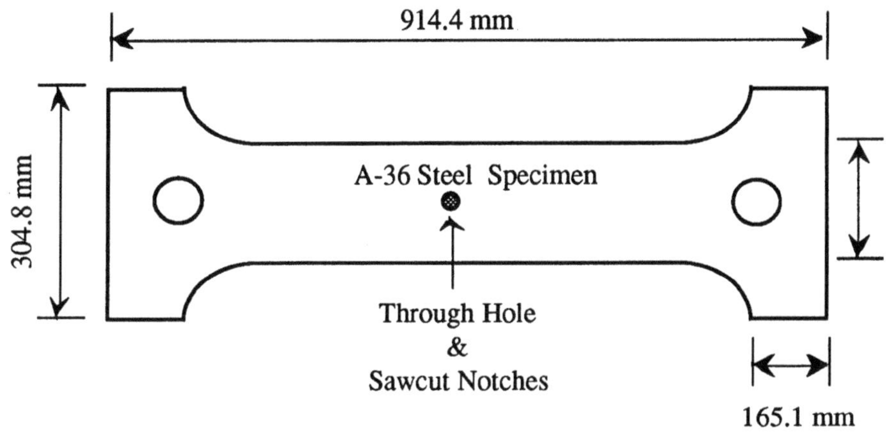

Fig. 4.70 The A-36 steel specimen used in the experiments.

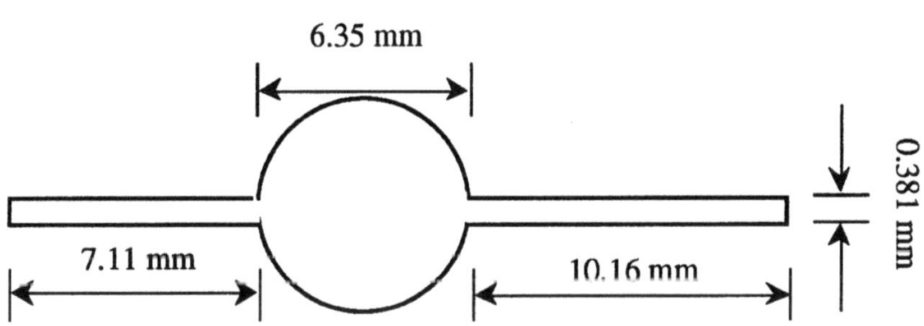

Fig. 4.71 Geometry of the hole and the starter notch.

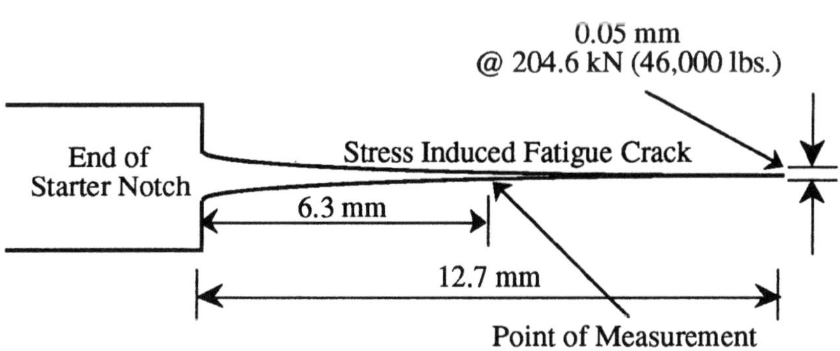

Fig. 4.72 Close-up geometry of the stress induced fatigue crack.

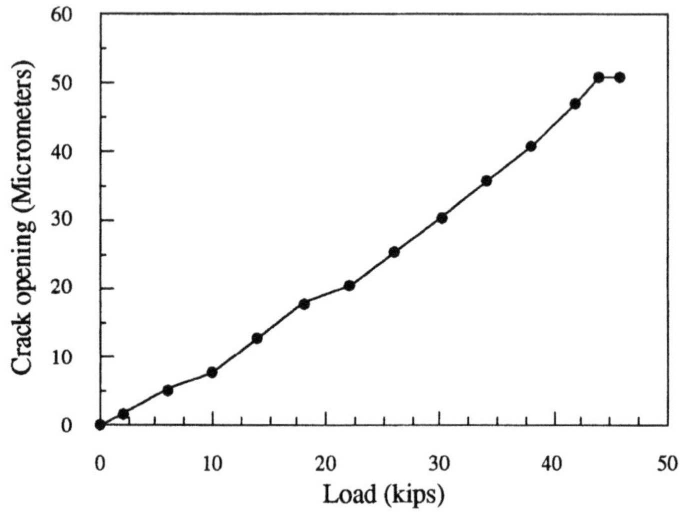

Fig. 4.73 Crack opening as a function of impressed loading.

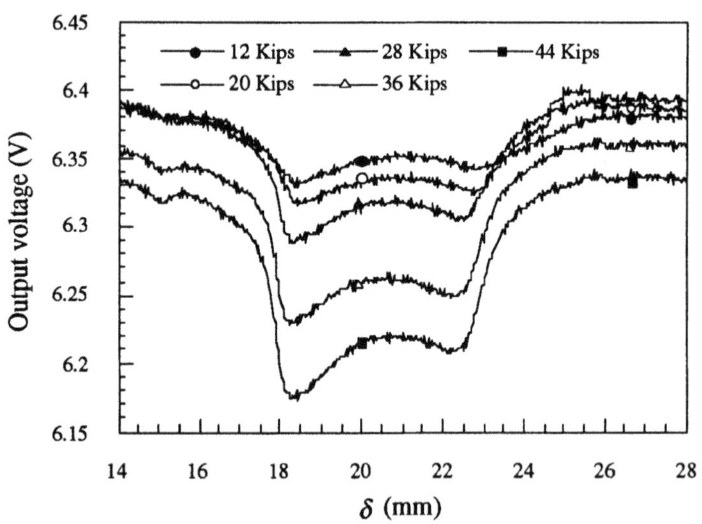

Fig. 4.74 Measured crack characteristic signals for the fatigue crack shown in Fig. 4.72 as a function of loading (or crack opening per Fig. 4.73).

- the method is fast, reliable and relatively inexpensive,

- the sensor may or may not be in contact with the surface under examination,
- cracks may be filled or covered with dielectric materials such as paint, dirt, rust, etc.,
- the same probe which detects and characterizes the properties of a crack under coating, may also (without any alteration to its design) measure the thickness of the coating and its material characteristics,
- cracks may be on non-ferromagnetic as well as ferromagnetic metals or alloys,
- cracks on the surface of graphite composites may also be detected and characterized,
- microwave techniques work with coarse-grained materials,
- the detected signal is only due to surface defects and not to interior flaws. Hence, signal interpretation is easier compared to techniques in which one must discriminate between signals due to interior and exterior flaws,
- the technique may be applied to curved and other complicated surfaces,
- the dimensions of a crack can be estimated, particularly its depth,
- crack orientation, edge and tip locations can be determined,
- no special operator skills, in the fields of microwaves or signal interpretation, are needed for successful crack detection,
- very little (if any) surface preparation is required. Any surface preparation which results in filling the crack with some particles may not affect detection sensitivity,
- the technique is environmentally compliant and operator friendly and safe,
- the required microwave power is in low milliwatt range, and does not create any safety related issues.
- a system may be battery operated and portable,
- the results are obtained in real-time,
- the technique is not a source of electromagnetic noise pollution (interference), and at the same time it is insensitive to external electromagnetic sources of interference. These two feature allow testing of a specimen during its normal operating conditions, thus reducing repair related down time,
- technique has the capability of inspecting large areas in a relatively short time (e.g. use of sensor arrays),
- defects in laminates and thick composites (disbond, void, inhomogeneity, impact damage, under cure, fiber bundle orientation and breakage, etc.) covering a metal specimen may also be detected and evaluated,
- these techniques are adaptable to automatic (no operator involvement) detection schemes.

It is expected that in the near future this microwave techniques and other similar to it will experience an increased level of practical exposure in many environments.

REFERENCES

Ash, E.A. and Husain, A. (1973) Surface examination using a superresolution scanning microwave microscope. *Proceedings of the 3rd European Microwave Conference*, pp. c.15.2.

Auld, B.A. (1978) Theory of ferromagnetic resonance probes for surface cracks in metals. G.L. Report 2839, E.L. Ginzton Laboratory, Stanford University, Stanford, CA.

Auld, B.A. (1981) Ferromagnetic resonance flaw detection. *Pys. Techniques*, **12**, pp. 149-154.

Bahr, A.J. (1980) Microwave eddy-current techniques for quantitative nondestructive evaluation. *Proceedings of DARPA/AF Review of Progress in Quantitative NDE*, La Jolla, CA.

Bahr, A.J. (1981) Microwave eddy-current techniques for quantitative nondestructive evaluation, in *Eddy-Current Characterization of Materials and Structures*, ASTM STP 722, (editors G. Birnbaum and G. Free) pp. 311-331.

Bahr, A.J. and Watjen, J.P. (1981) Novel eddy-current probe development. Semi-annual Report, Contract F33615-80-C-5025, SRI Project 1908, Menlo Park, CA.

Bakhtiari, S., Ganchev, S.I., Qaddoumi, N. *et al.*, (1994) Microwave non-contact examination of disbond and thickness variation in stratified composite media. *IEEE Transactions on Microwave Theory and Techniques*, **42** (3), pp. 389-95.

Borgnis, F.E. and Papas C.H. (1958) Encyclopedia of Physics: Electromagnetic Waveguides and Resonators, **16**, Springer Verlag.

Boving, K.G. (1989) NDE handbook, non-destructive examination methods for condition monitoring. *Teknisk Forlag A/S (Danish Technical Press).*

Collin, R. E. (1992) Foundations for microwave engineering, McGraw-Hill, New York.

Feinstein, L. and Hruby, R.J. (1967) Surface crack detection by microwave methods. *6th Symp. on Nondestructive Evaluation of Aerospace and Weapons Systems Components and Materials*, San Antonio, Texas.

Feinstein, L. and Hruby, R.J. (1968) *AIAA/ASME 9th Structure, Structural Dynamics and Materials Conference*, paper #68-321.

Ganchev, S.I., Zoughi, R., Huber, C. *et al.*, (1996) Microwave method for locating surface slot/crack tips in metals, *Materials Evaluation*, **54** (5), pp. 598-603.

Gysel, U.H. and Feinstein, L. (1974) Design and fabrication of stripline microwave surface-crack detector in projectiles. Final report, Contract DAAG-46-73-C-0257, SRI Project 2821, Stanford Research Institute, Menlo Park, CA.

Harrington, R. F. (1961) Time harmonic electromagnetic fields, McGraw-Hill, New York.

Harrington, R.F. and Mautz, J.R. (1976) A generalized network formulation for aperture problems. *IEEE Transactions on Antennas Propagation*, **AP-24** (6), pp. 870-73.

Harrington, R.F., (1982) Field computation by moment methods, Krieger Publishing Co.

Hruby, R.J. and Feinstein, L. (1970) A novel non-destructive non-contacting method for measuring the depth of thin slits and cracks in metals. *Review of Scientific Instruments*, **41**, pp. 679-683.

Huber, C. (1996) Electromagnetic modeling of exposed and covered surface crack detection using open-ended waveguides. Ph.D. Dissertation, Electrical Engineering Department, Colorado State University.

Huber, C., Abiri, H., Ganchev, S.I. *et al.*, (1997a) Analysis of the "crack characteristic signal" using a generalized scattering matrix representation. *IEEE Transactions on Microwave Theory and Techniques*, **45** (4), pp. 477-84.

Huber, C., Ganchev, S.I., Mirshahi, R. *et al.*, (1997b) Remote detection of surface cracks/slots using open-ended rectangular waveguide sensors: An experimental investigation. *Nondestructive Testing & Evaluation*, vol. 13, pp. 227-237.

Husain, A. and Ash, E.A. (1975) Microwave scanning microscopy for nondestructive testing. *Proc. of the 5th European Microwave Conference*, Hamburg, Germany, pp. 213-217.

Marcuvitz, N. (1951) Waveguide handbook, McGraw-Hill, New York.

Pozar, D. (1990) Microwave engineering, Chapter 4, Addison Wesley, New York.

Qaddoumi, N., Mirshahi, R., Ranu, E. *et al.*, (1998a) Analysis of loading vs. microwave fatigue crack detection sensitivity using open-ended waveguides. *Proceedings of the Eighth International Symposium on Nondestructive Characterization of Materials VIII*, pp. 311-16, Boulder, CO.

Qaddoumi, N., Ranu, E., Mirshahi, R. *et al.*, (1998b) Development of equipment for and quantitative analysis of microwave detection and evaluation of fatigue induced surface cracks in steel, *Final Report*,

Federal Highway Administration (FHWA), Grant no. DTFH61-94-X-00023, p. 519, January.

Robinson, L.A. and Gysel, U.H. (1972) Microwave coupled surface crack detector. Final report, Contract DAAG-46-72-C-0019, SRI Project 1490, Stanford Research Institute, Menlo Park, CA.

Uher, J., Bornemann, J. and Rosenberg, U. (1993) Waveguide components for antenna feed systems: Theory and CAD, Artech House, Boston.

Yeh, C. (1994) Detection and sizing of surface cracks in metals using open-ended rectangular waveguides. Ph.D. Dissertation, Electrical Engineering Department, Colorado State University.

Yeh, C. and Zoughi, R. (1994a) A novel microwave method for detection of long surface cracks in metals. *IEEE Transactions on Instrumentation and Measurement*, **43** (5), pp. 719-25.

Yeh, C. and Zoughi, R. (1994b) Microwave detection of finite surface cracks in metals using rectangular waveguide sensors. *Research in Nondestructive Evaluation*, **6** (1), pp. 35-55.

Yeh, C. and Zoughi, R. (1995) Sizing technique for slots and surface cracks in metals. *Materials Evaluation*, **53** (4), pp. 496-501.

Yeh, C., Ranu, E. and Zoughi, R. (1994) A novel microwave Method for surface crack detection using higher order waveguide modes. *Materials Evaluation*, **52** (6), pp. 676-681.

Zoughi, R., Ganchev, S.I. and Huber, C. (1996) Measurement parameter optimization for surface crack detection in metals using an open-ended waveguide probe," *Proceedings of the IEEE Instrumentation & Measurement Technology Conference*, pp. 1391-1394, Brussels, Belgium, June 4-6.

Zoughi, R., Ganchev, S.I., Huber, C. *et al.*, (1995) A novel microwave method for filled and covered surface crack detection in steel bridge members including crack tip identification. *Fourth Quarterly Report*, Contract DTFH61-94-X-00023, Federal Highway Administration.

CHAPTER 5

Near-field measurement techniques and applications

5.1 INTRODUCTION

In Chapters 3 and 4 open-ended rectangular waveguides were used to inspect multi-layered dielectric composites and metals for surface cracks, respectively. Microwave non-destructive measurements are not exclusively conducted using open-ended rectangular waveguide probes. As mentioned in Chapter 1 other probes such as open-ended coaxial lines, monopole antennas, stripline resonators and many others may be employed depending on the application. However, no matter which probe is used, there is still a need for a complete measurement system complementing the probe used. Such a system requires a microwave frequency generator, some type of a detector and a means to separate the incident signal from that reflected by the specimen under test, or a mechanism to combine these two signals. Furthermore, unlike far-field approaches (i.e. radar systems) there is a wide range of parameters of interest that may need to be measured. These parameters include phase and magnitude of reflection coefficient, phase and magnitude of transmission coefficient, reflected and transmitted power, standing wave properties, quality factor and resonant frequency of a cavity, just to name a few. Consequently, there exists a wide variety of measurement systems that are capable of providing these sought for parameters.

In addition, there are other requirements that may influence the choice of the measurement system. For example, if the goal of an investigation is to verify the results of a theoretical investigation, then one needs to conduct measurements whose outcome is referenced to the same reference point in the measurement system as in the theoretical derivation. In Chapter 3 this information was the phase and magnitude of reflection coefficient calculated at the aperture of an open-ended rectangular waveguide probe. Consequently, the measurements were conducted in a fashion to provide the

---

* *The photos reproduced here in black&white are reproduced in FULL-COLOUR in the front of this book.*

same information referenced to the waveguide aperture. In Chapter 4 the parameters of interest, calculated using the derivations, were the standing wave pattern properties inside the probing waveguide. To verify the calculated results, the measurement system was designed to provide the same information. The type of measurements mentioned thus far may be considered in the category of calibrated measurements. These measurements can be conducted using expensive and state-of-the-art microwave vector network analyzers, or using a custom-designed measurement system which may then be calibrated using several calibration loads such as a short circuit, a matched load and a known load. In recent years, several investigators have developed improved calibration techniques which do not require extremely well designed and well defined calibration loads (Ganchev *et al.*, 1995a).

However, in many cases only relative information indicating the presence of a defect is needed. For example, when conducting near-field inspection of composite structures, the primary objective is to detect the presence of a defect. The secondary objective is ascertaining the properties of the defect such as its spatial extent, depth at which it occurs within the composite, etc. These measurements can be conducted using uncalibrated measurement systems.

In this chapter the design of basic microwave measurement systems for near-field non-destructive testing applications will be presented. Subsequently, several important and wide ranging examples of near-field measurements and imaging will also be presented.

5.2 MEASUREMENT TECHNIQUES

When conducting near-field microwave measurements, and depending on the measurement parameter of interest, there may exist a wide range of techniques that may be applicable. As mentioned in the previous section, there are two distinct measurement techniques; namely, calibrated and uncalibrated techniques. The calibrated technique is usually used when verifying theoretical results or performing inverse problems, while the uncalibrated techniques are more often used in practical situations in which defect detection is the primary goal.

5.2.1 Calibrated microwave measurement techniques

Calibrated in this context refers to measurements that are referenced to a specific measurement plane within a system. Usually, this reference plane is the aperture of an open-ended probe such as an open-ended rectangular waveguide or a coaxial probe. As it was demonstrated in Chapter 3, when calculating the properties of reflection (or transmission) coefficient, the ratio of the reflected to the incident waves at the waveguide aperture is obtained. There are two basic approaches in making calibrated microwave measurements. The first approach involves the use of vector network analyzers. These network analyzers are capable of producing robust reflection and transmission coefficient information with high accuracy. This is accomplished by considering a measurement setup as a multi-port microwave device whose scattering parameters are subsequently measured by the network analyzer. Different scattering parameters associated with the scattering matrix of the equivalent multi-port device (e.g. the measurement system) represent reflection and transmission coefficient characteristics of the device.

Figure 5.1 shows the schematic of a measurement setup for measuring the reflection coefficient of an open-ended probe radiating into a conductor-backed multi-layered composite structure. This represents a single-port measurement for which only the phase and magnitude of the reflection coefficient, referenced at the aperture of the open-ended probe, are sought as a function of the composite layered structure, dielectric properties, the frequency of operation and the standoff distance. To obtain the reflection coefficient properties one needs to calibrate the network analyzer. This requires that the open-ended probe be terminated in several known loads such as a short circuit (moving short in some cases), a matched load (moving load in some cases), and a known load (leaving the open-ended probe radiating into free-space may satisfy this requirement for some applications). Once the reflection properties of these loads have been measured and taken into account by the network analyzer, the reflection coefficient of an unknown load (e.g. the multi-layered composite structure) may be measured, and subsequently related to the geometrical and dielectric properties of the composite structure (for a given frequency of operation). This type of calibration may also be conducted over a relatively wide range of frequencies depending on the capabilities of the network analyzer.

Figure 5.2 shows a two-port measurement setup for evaluating the dielectric properties of a specimen cut-to-fit inside a section of a waveguide sample holder (as explained in Chapter 2). In this case the properties of the reflected as well as the transmitted microwave signals are used to determine the relative permittivity and loss factor of the dielectric specimen. In this case these properties are measured once when the sample holder is empty and once when it is filled with the dielectric specimen. The measured change in the reflection and transmission properties of the microwave signals between these two cases is used to obtain the dielectric properties of the specimen (Baker-Jarvis, 1990; Weir, 1970).

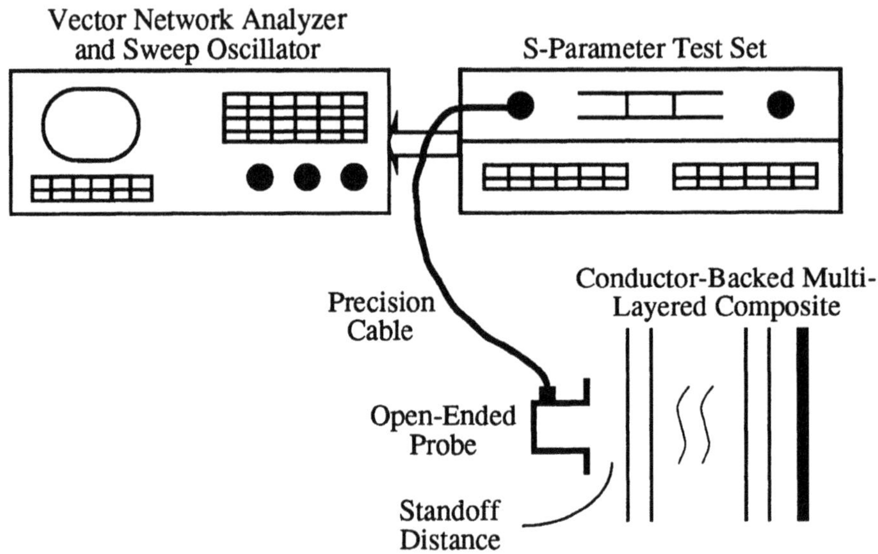

Fig. 5.1 A single-port calibrated measurement system utilizing a vector network analyzer.

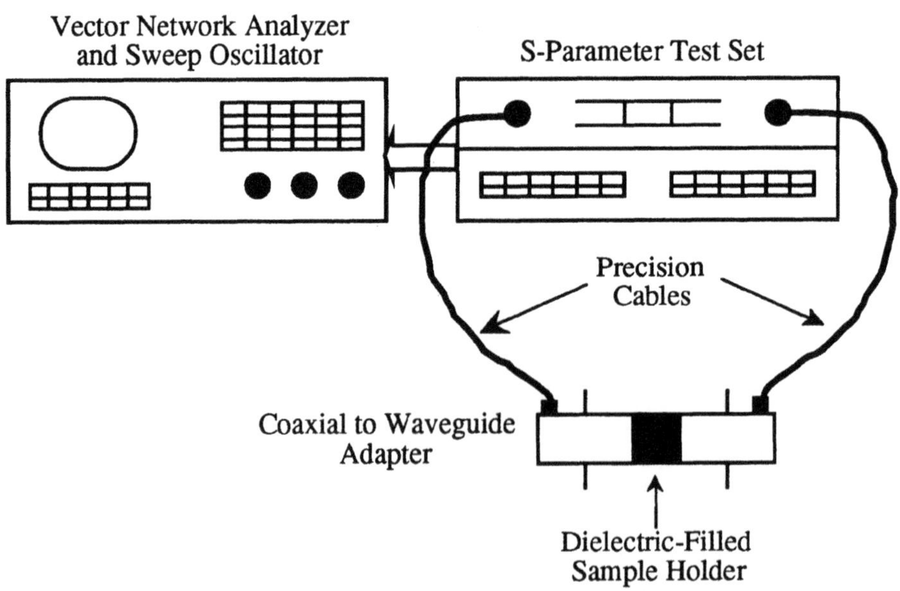

Fig. 5.2 A two-port calibrated measurement system utilizing a vector network analyzer.

Although vector network analyzers provide for a robust and accurate measurement of the reflected and transmitted microwave signal properties, it is not required to use them exclusively when making calibrated measurements. Custom-designed microwave measurement systems can also be calibrated to give referenced information about the reflection and transmission coefficient characteristics. In recent years several investigators have attempted to develop robust calibration schemes for this purpose. In particular such schemes have been developed when using open-ended coaxial probes since it is much more difficult to obtain known calibration loads for this probe than for an open-ended rectangular waveguide probe (Ganchev *et al.*, 1995a; Stuchly and Stuchly, 1980; Mosig *et al.*, 1981; Kraszewski, Stuchly and Stuchly, 1983; Nyshadham, Sibbald and Stuchly, 1992; Liping, Deming and Zhiyan, 1986; Jenkins, Warham and Clarke, 1992; Bakhtiari, Ganchev and Zoughi, 1993).

When using custom-designed measurement systems which subsequently need to be calibrated, one drawback is that one may have to characterize each microwave component that is used in the system. Then, it is possible to obtain the flow graph of the system and evaluate its properties using standard microwave network analysis (Pozar, 1998). In such cases repeated calibrations may be required, particularly if a connection between two components changes. This may be time consuming for some applications. In addition, good calibration can only be obtained at a single frequency and swept frequency calibration is not easily attained. Another calibration approach for such measurement systems is to consider the entire system, starting from its input and ending at its output ports, as an equivalent two-port device (Ganchev *et al.*, 1995a). In this way one does not need to know the scattering properties of each individual microwave component. Subsequently, by using enough number of known loads the system may be calibrated. Once again, good calibration is usually obtained only for a narrow range of frequencies.

5.2.2 Uncalibrated microwave measurement techniques

This type of measurements does not generally require any calibration since the goal of these measurements is not to determine absolute information about the parameter of interest nor is it to perform a rigorous inverse problem. Uncalibrated measurement techniques are generally associated with applications in which relative information about the presence and properties of a defect are sought. It must be mentioned here that a defect must first be detected before its geometrical or other properties can be evaluated. For example, when inspecting a thick multi-layered composite, for the presence of disbonds between its various layers, one must first ascertain the presence of the disbond. Once this is accomplished then disbond properties (such as its depth as discussed in Chapter 3) may be evaluated.

There is a wide variety of near-field microwave measurement systems that can be custom designed, built and effectively used. The particular microwave design related features associated with these systems depend on the measured parameter of interest. For example, when making near-field, non-contact and one-sided measurements of most low-loss dielectric materials, including the type of thick sandwich composites mentioned in Chapter 3, the presence of a lossless or low-loss defect, such as a disbond, is often detected using the phase of reflection coefficient. Therefore, for the purpose of detection, it is sufficient to utilize a measurement apparatus whose output is primarily sensitive to the phase of reflection coefficient (i.e. a phase detector). On the other hand, when inspecting the presence of lossy anomalies, such as moisture permeation or rust under paint, within a low-loss dielectric composite, the magnitude of reflection coefficient may be used to indicate the presence of the defect. In such cases, a simple reflectometer circuit may be used whose output is primarily a function of the magnitude of the reflection coefficient. It is important to note that when making the type of measurements mentioned here, it is difficult to obtain results that are entirely dependent upon the phase or magnitude of a signal (i.e. reflected signal or reflection coefficient). This is due to the fact that microwave components that are usually used in the design of such circuits are not ideal and there will always be some unwanted signal leakage, reflections, feed-throughs, etc. which contaminate the output signal. This is why the word "primarily" was used earlier.

In some uncalibrated measurements involving the use of a slotted line or a detector fixed at a location within a transmission line in which both the transmitted and reflected signals are present (such as the case described in Chapters 2 and 4) the measured output voltage is not only a function of the magnitude of the reflected signal but also its phase in addition to the detector input power-output voltage characteristics. This is why a calibrated rotary-vane attenuator was incorporated in the slotted line measurement apparatus described in Chapter 2. Therefore, when making this type of measurements one must be careful to interpret the results correctly.

Figure 5.3 shows a general schematic of a near-field, non-contact and one-sided measurement system. An oscillator feeds an open-ended sensor or probe via a directional signal splitter/divider. This is done so that a portion of the transmitted signal is available to be compared to or mixed with the reflected signal. The main portion of the signal, supplied by the oscillator, is then transmitted out of the sensor which becomes incident upon the specimen under inspection (such as the composite specimen shown).

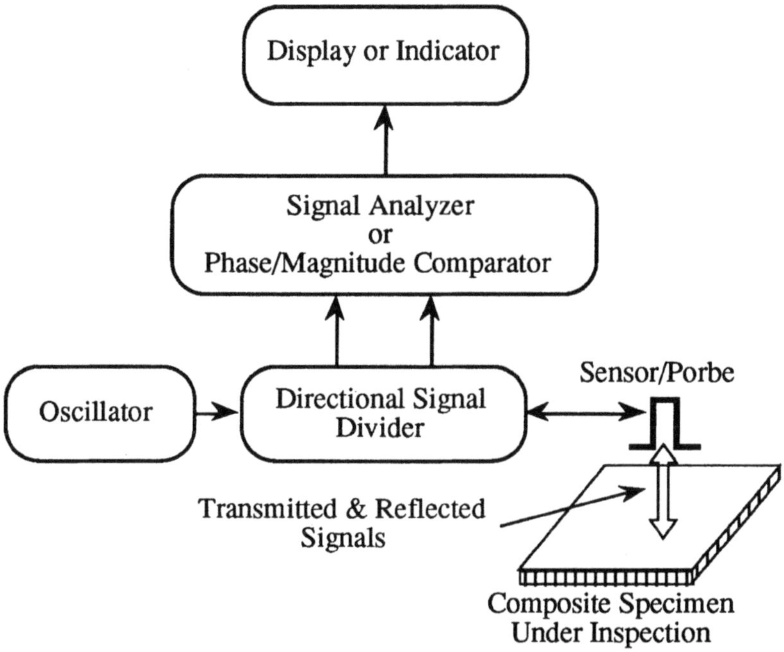

Fig. 5.3 General schematic of a near-field, non-contact and one-sided (reflection) measurement system.

The incident signal interacts with the specimen under inspection and a portion of it is reflected back towards the open-ended probe which receives it. This reflected signal enters into the signal analyzer or the phase/magnitude comparator (i.e. network analyzer) via the directional signal divider. The phase and/or magnitude of these two signals are compared and a signal proportional to their difference is fed into a display or indicator. Figures 5.4 and 5.5 illustrate general designs for a phase and a magnitude detector, respectively. It is useful for the reader to consult with (Gardiol, 1984; Pozar, 1998) for more detailed and comprehensive use of each of the particular microwave components and devices shown in these figures.

As the specimen under inspection and the open-ended sensor move with respect to each other (i.e. raster or C-scan) the reflected signal from the specimen changes if an anomaly, moisture contamination, an inclusion, a disbond, a void or a flaw is present. The change in the phase and/or magnitude of the reflected signal due to any of the mentioned defects once compared with the reference signal in the signal analyzer indicates the presence of the defect. This change may also be used to obtain information about the spatial extant, the type of defect encountered and its size.

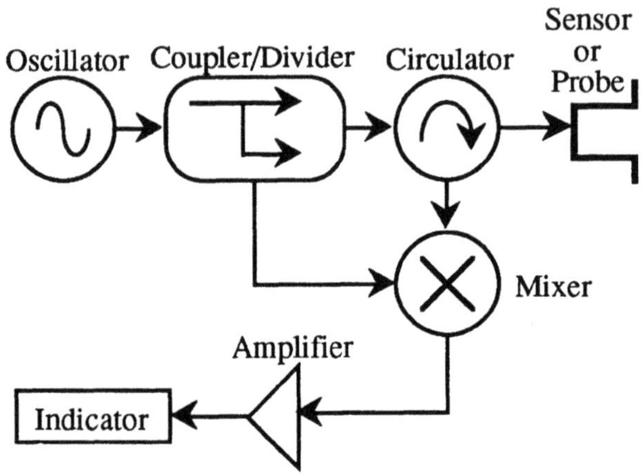

Fig. 5.4 General schematic of a phase detector.

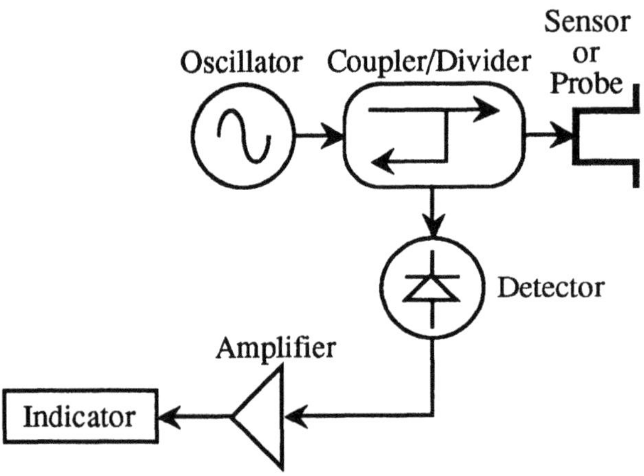

Fig. 5.5 General schematic of an amplitude detector.

Besides the standoff distance and the frequency of operation, one may use the characteristics of various microwave components and devices used in a given measurement system to improve its sensitivity to the presence of a defect. For example, extra electrical lengths strategically added to different sections of a phase sensitive system may bias the system in such a way to yield more sensitive results to slight phase changes. Therefore, the knowledge of optimizing measurement system hardware design should be considered an important issue when designing microwave inspection systems. It must also be mentioned here that more complex and sophisticated measurement systems employing in-phase and quadrature (I-Q) output channels may also be designed and used (King and Yen, 1981; Bolomey and Pichot, 1990). However, the complexity of such a system may not provide a distinct advantage when used for most near-field applications.

5.3 MEASUREMENT PROCEDURES

There are numerous applications for which a calibrated or an uncalibrated near-field microwave measurement technique may be employed. These measurements typically involve measuring an output voltage, which is primarily proportional to the phase or magnitude of reflection or transmission coefficients, as function of frequency and standoff distance over a specimen under test (similar to Fig. 5.3). As explained in Chapter 3, standoff distance and frequency optimization is an important issue in many applications. Therefore, it is important to briefly discuss how this optimization is accomplished before near-field microwave imaging is discussed.

5.3.1 Frequency and standoff distance optimization[1]

In a majority of cases for which a custom-designed measurement system is sought, a calibration target or specimen is used to aid in designing the system. A calibration target or specimen usually consists of a specimen similar to the one that is of interest in real life except with some known defects embedded in it. Assuming such a specimen is available, then one may find the (near) optimum frequency and standoff distance for detecting the presence of the defect. Optimum frequency and standoff distance provide for the most robust detection of the defect for a given measurement system.

Optimum frequency of operation may be determined either using analytical solutions, such as that presented in Chapter 3, or it may be found experimentally. Consider that a measurement system has been designed and constructed utilizing an open-ended sensor. In addition, let us assume that the microwave frequency band for a given application has been determined. Such systems and their components usually operate in a range of frequencies (i.e. the waveguide band). Optimization may then be necessary in order to determine the optimum operating frequency within

---

[1] Portions of this sub-section are reprinted with permission from *Materials Evaluation*, vol. 53, no. 8, pp. 926-29, Copyright 1995 © The American Society for Nondestructive Testing, Inc.

this range. Since this is a near-field measurement, the influence of standoff distance must also be considered. Consequently, unlike the far-field approach there are two measurement parameters that must be simultaneously determined. Therefore, commonly the following procedure is followed:

1. The sensor is placed over a region devoid of defects and flush with the specimen (i.e. in-contact). The frequency is then varied within the operating frequency range while recording the measured output signal or voltage. The sensor is then placed over the defective region and the same process is repeated.
2. The frequency at which the difference between the two measured voltages are maximum is the optimum frequency for the in-contact case.
3. The same process is then repeated for other standoff distances.
4. Subsequently, the standoff distance and frequency of operation at which the largest voltage difference is obtained are the optimum measurement parameters.

This is a simple description of the experimental optimization process. There are other issues such as measured output voltage sensitivity to slight standoff distance or frequency variations that may also need to be considered.

To show the importance of standoff distance analysis for enhanced defect detection purposes an experiment was conducted (Qaddoumi *et al.*, 1995a). A panel, as shown in Fig. 5.6, was made of a 25.4 mm-thick glass reinforced polymer epoxy. This panel had an aluminum inclusion of 6.35 mm by 6.35 mm by 0.8 mm located at a distance of 12.7 mm from the surface of the panel. A phase sensitive measurement apparatus, operating at a frequency of 10.5 GHz and employing an open-ended rectangular waveguide aperture probe, was used to investigate the potential of detecting the aluminum inclusion, and to demonstrate the detection sensitivity to standoff distance variation. The frequency of 10.5 GHz had previously been shown to be a sensitive frequency for detecting aluminum inclusions in this panel. It must be mentioned that it is expected that an aluminum inclusion is easily detected using microwave techniques, since aluminum is a perfect conductor and strongly reflects the microwave signal. Therefore, the goal of this measurement was primarily academic and for demonstration. Two measurements were performed in which the standoff distance between the probe and the surface of the sample was changed. In the first measurement the probe was placed directly on the region in which the inclusion existed, and in the second measurement it was placed on a region devoid of the inclusion. Subsequently, the standoff distance was

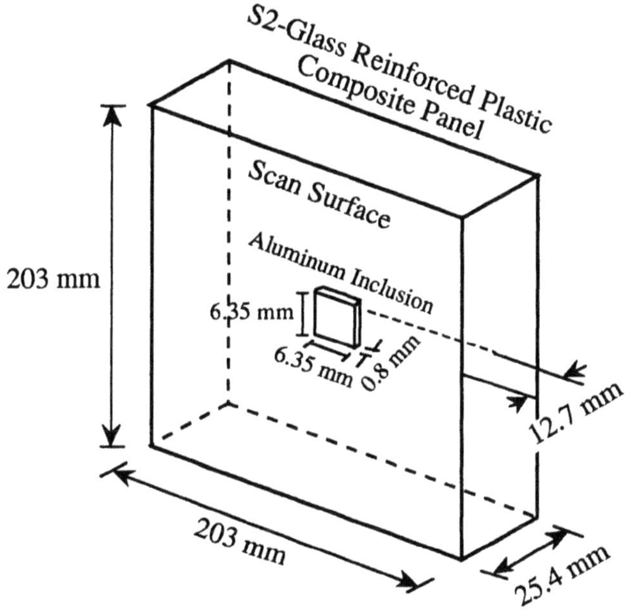

Fig. 5.6 Schematic of a thick glass reinforced polymer epoxy with an aluminum inclusion (Qaddoumi *et al.*, 1995a).

increased, and the voltage was recorded, as shown in Fig. 5.7 (Qaddoumi *et al.*, 1995a). The results show that operating at a standoff distance of 0 mm (i.e. in-contact) and at this frequency (10.5 GHz) the difference between the signals, corresponding to the regions with and without the aluminum inclusion, is maximum. On the other hand, if a standoff distance of around 5 mm or 13 mm is used, there will be no distinction between these two regions. This is due to the coherent interference among signals reflected from the front surface of the panel, the aluminum inclusion and the back face of the panel. It is important to understand that in the near-field approach, similar to the far-field approach, the reflected signal reaching the waveguide aperture is a coherent combination of reflected signals from various boundaries within the structure (see Chapter 3). Therefore, it is expected that for a given situation, depending on the geometrical and material characteristics of the specimen and the inclusion, the standoff distance and the frequency of operation even a strongly reflective inclusion, in this case aluminum, may not be detected. Another observation is that, operating in the standoff distance range of 5-13 mm the contrast is reversed when compared to the in-contact case. Moreover, in this region the difference between the voltages is about one third of that for the in-contact

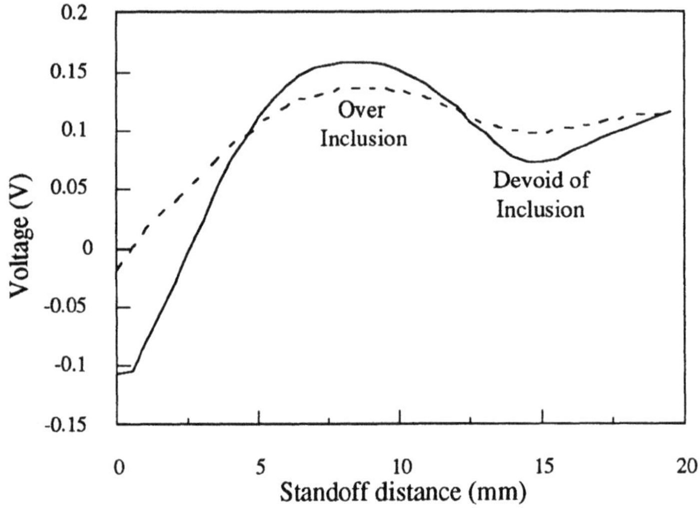

Fig. 5.7 Recorded voltage over the inclusion and over an area devoid of the inclusion as a function of standoff distance at a frequency of 10.5 GHz (Qaddoumi *et al.*, 1995a).

case resulting in a relatively less sensitive detection. Also, operating in the range where the output as a function of the standoff distance is almost constant (7-10 mm) is important from a practical point of view since slight variations in the standoff distance do not significantly influence the outcome. This example clearly demonstrates the effectiveness of using standoff distance analysis for increased defect detection sensitivity.

5.4 NEAR-FIELD IMAGING

As mentioned earlier, microwave imaging is based on transmitting a wave onto a material and using the magnitude and/or phase information of the transmitted and/or the reflected signals to create a two or three dimensional image of the material (d'Ambrosio, *et al.*, 1993; Bolomey and Pichot, 1990; Gopalsami *et al.*, 1994). Transmission type microwave approaches require access to both sides of the sample. To achieve high spatial resolution (detection of small defects) very high frequency transmission techniques have been used (Gopalsami *et al.*, 1994). However, as will be seen in the following sections the utilization of high frequencies is not usually necessary for achieving adequately high spatial resolution. This is a direct result of the fact that when operating in the near-field of a probe the spatial resolution is primarily influenced by the probe dimensions. Probe dimensions at microwave and millimeter wave frequencies are relatively small resulting in relatively high spatial resolution measurements and images.

5.4.1 Inclusions in glass reinforced polymer epoxy[2]

To demonstrate the type of images and features that can be obtained using near-field microwave imaging the panel shown in Fig. 5.6 was used, in conjunction with the measurement schematic described in Figs. 5.4 and 5.5, to produce near-field images of the aluminum inclusion at a frequency of 10.5 GHz at two standoff distances of 0 mm (i.e. open-ended waveguide probe in contact with the panel) and 9 mm, respectively. These two standoff distances correspond to two important regions of the standoff distance curve shown in Fig. 5.7. The horizontal and vertical scan steps were 0.5 mm and 2 mm, respectively. The scan covered an area of 85 mm by 98 mm. The recorded data was then used to generate a three-dimensional contrast image consisting of the two spatial scanning dimensions and the intensity associated with each data point comprising the third dimension. Upon having generated a two-dimensional matrix of the measured data, they are normalized with respect to the largest measured voltage. Subsequently, various shades of color are assigned to each normalized data point and a normalized color (false) image is generated. It is important to note that each image shown in this chapter is referenced (normalized) to itself. The colors associated with one image do not correlate with or correspond to the same colors in another image.

Figure 5.8 shows the measured signal intensity for the in-contact case. This data was then used to produce an image, as shown in Fig. 5.9 (Qaddoumi *et al.*, 1995a). The results indicate a high contrast, and the aluminum inclusion is clearly visible. This is as a direct result of the standoff distance analysis for this specimen (see Fig. 5.7 and zero standoff distance). Both figures also show a very good indication of the fiber bundle pattern associated with the opposite side of the sample. Also, the indication of these bundles is more pronounced in the vertical direction than the horizontal direction. This is due to the fact that the polarization of the sensor was aligned with the vertically oriented fiber bundles. This fact also demonstrates the potential of using polarization diversity for certain applications in which the preferred orientation of a defect is of interest (Zoughi, Cone and Nowak, 1991; Ulaby, Moore and Fung, 1986; and also see Chapter 4).

---

[2] Portions of this sub-section are reprinted with permission from *Materials Evaluation*, vol. 53, no. 8, pp. 926-29, Copyright 1995 © The American Society for Nondestructive Testing, Inc.

Fig. 5.8 A phase scan intensity image of the composite panel shown in Fig. 5.6 at a frequency of 10.5 GHz and when in contact (Qaddoumi *et al.*, 1995a).

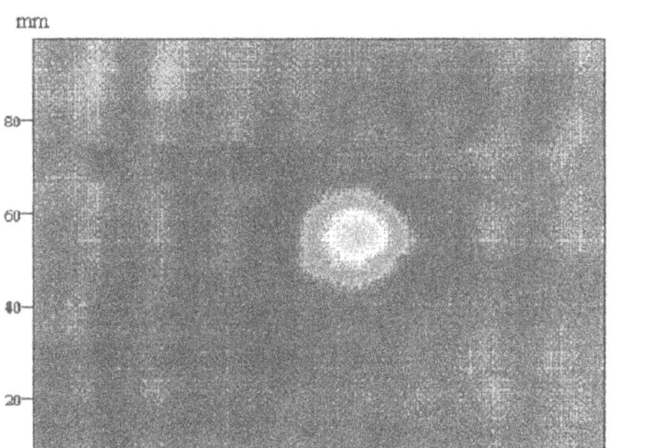

Fig. 5.9 A phase scan image of the composite panel shown in Fig. 5.6 at a frequency of 10.5 GHz and when in contact (Qaddoumi *et al.*, 1995a).

Figures 5.10 and 5.11 show similar results obtained at 9 mm-standoff distance. The results show that the contrast is reduced (i.e. the dynamic range is less) and reversed (compared to the results of Fig. 5.8). However, at this standoff distance, the fiber bundle orientation seems to be more visible than that in Fig. 5.9. This is another indication of the sensitivity of the standoff distance to different specimen characteristics (e.g. small thickness variation due to the bulging of the fiber bundles on the opposite face of the panel).

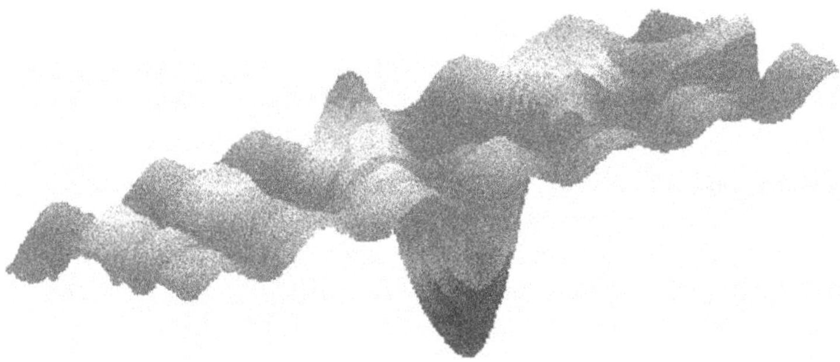

Fig. 5.10 A phase scan intensity image of the composite panel shown in Fig. 5.6 at a frequency of 10.5 GHz and at a standoff distance of 9 mm (Qaddoumi *et al.*, 1995a).

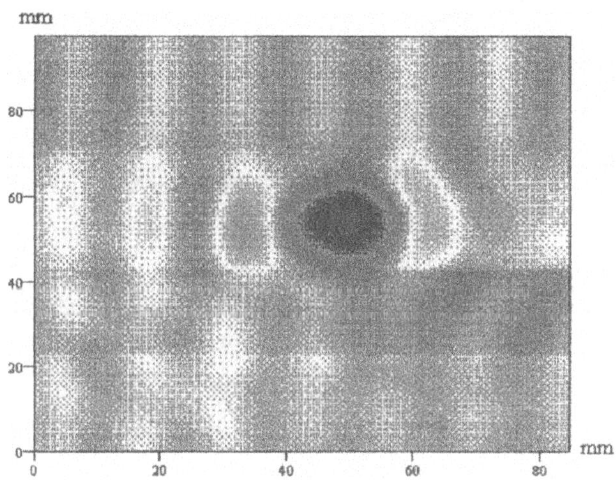

Fig. 5.11 A phase scan image of the composite panel shown in Fig. 5.6 at a frequency of 10.5 GHz and at a standoff distance of 9 mm (Qaddoumi *et al.*, 1995a).

To demonstrate the potential of using such images for anomaly size discrimination, another similar panel was produced, and a section of it was used for this purpose. The panel was a 38 mm-thick glass reinforced polymer composite with three embedded aluminum inclusions of 25.4 mm by 25.4 mm, 12.7 mm by 12.7 mm and 6.35 mm by 6.35 mm, (all 0.8 mm thick) all at a depth of 19 mm from the surface. This panel was also scanned at a frequency of 10.5 GHz, with the waveguide sensor in contact with the panel surface (Qaddoumi *et al.*, 1995a). Figure 5.12 shows the image of an amplitude scan (i.e. the voltage measured is related to the amplitude of the reflected signal at the aperture of the waveguide). The scanned area was 150 mm by 58 mm. The horizontal and the vertical scan steps were 0.5 mm and 2 mm, respectively. Using these raw data without any signal/image processing enhancement, and without any rigorous optimization of the measurement parameters in this case (frequency and standoff distance), all three defects are distinctly detected. A strong indication of the relative size of each defect is also evident. The measured distances between the centers of the defects obtained from the image correspond well with the actual distances.

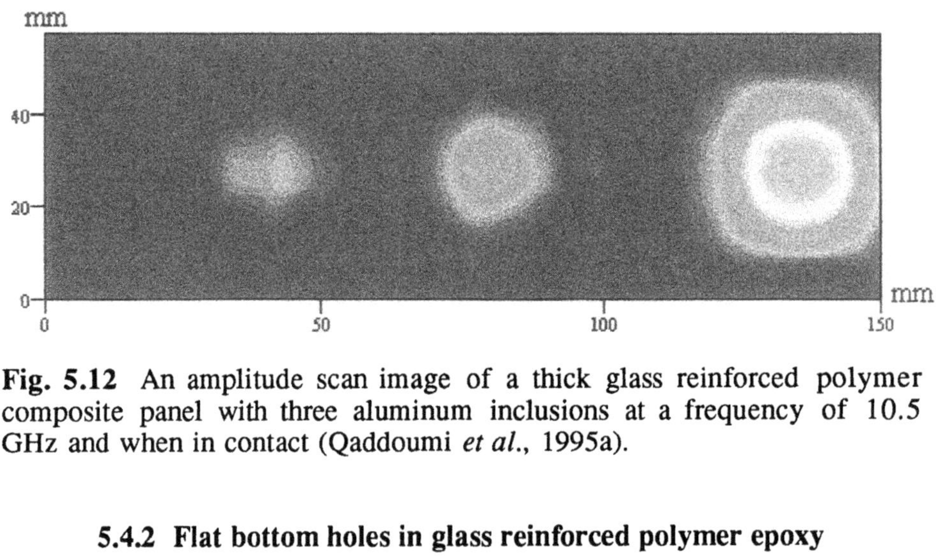

Fig. 5.12 An amplitude scan image of a thick glass reinforced polymer composite panel with three aluminum inclusions at a frequency of 10.5 GHz and when in contact (Qaddoumi *et al.*, 1995a).

5.4.2 Flat bottom holes in glass reinforced polymer epoxy

Flat bottom holes are often used as standard flaws to gain an understanding of the potential of a non-destructive inspection technique for disbond, delamination and void detection. Subsequently, a panel made of a 38 mm-thick glass reinforced polymer composite was manufactured with several flat bottom holes. The area scanned from this sample had two flat bottom holes with diameters of 6.35 mm and at 19 mm and 12.7 mm from the surface of the sample, as depicted in Fig. 5.13 (Qaddoumi *et al.*, 1995a). The scanned area was 100 mm by 22 mm. The horizontal and the vertical scan steps were 0.5 mm and 2 mm, respectively. A scan of this sample was conducted in-contact at a frequency of 24 GHz since this frequency showed to be more sensitive to the presence of these holes (additionally at higher frequencies the waveguide aperture is smaller and hence the measurements are more sensitive, spatially, to the presence of embedded flaws). The image, depicted in Fig. 5.14, shows a clear indication of both holes (voids) and the signal strength in each case may be related to the depth of each hole. The measured distance between the two hole centers on the image closely matches the actual distance between the centers on the panel. Two adjacent regions of relative high intensity are seen on each side of the holes. They are attributed to the hole walls, and the interaction between the hole side walls and the radiation pattern of the open-ended rectangular waveguide probe (Qaddoumi *et al.*, 1995a; Qaddoumi and Zoughi, 1997). It is important to note that other non-destructive techniques that had been used to study this particular specimen (ultrasonics and thermal imaging) did not detect these holes (Carriveau, 1994). Difficulty with ultrasonics and thermal imaging was found to be a result of the interference caused by a very large area disbond in near surface laminates. This additional unintentional masking defect did not interfere with the microwave measurements because microwaves could penetrate through the disbond and interrogate the following layer. It is important to note that the detection of near surface disbonds and voids with microwaves is also possible with some tuning of the measurement parameters such that the measurement results would be sensitive to the presence of disbonded layers (see Chapter 3). A larger scan of this area, using the same microwave technique, was conducted which in addition to the holes showed the near surface disbonded region as well (when compared with non-disbonded regions).

To improve the image contrast (difference between the signals due to the holes and the background), other amplitude scans at several standoff distances were conducted. Figure 5.15 shows an image of the same panel at 24 GHz and at a standoff distance of 0.5 mm. The indications of the holes on this image are much stronger than those shown in Figure 5.14 (i.e. higher contrast). This illustrates the importance of the choice of standoff distance. This is an example of how a measurement may be optimized for contrast enhancement. As discussed in Chapter 3, the frequency of operation and the standoff distance may be tuned to increase measurement sensitivity including sensitivity to the presence of a defect.

5.4.3 Disbond in thick sandwich composite[3]

As explained in Chapter 3, one of the more likely occurring flaws in thick

[3] Portions of the next two sub-sections are reprinted with permission from *Materials Evaluation*, vol. 53, no. 4, pp. 463-67, Copyright 1995 © The American Society for Nondestructive Testing, Inc.

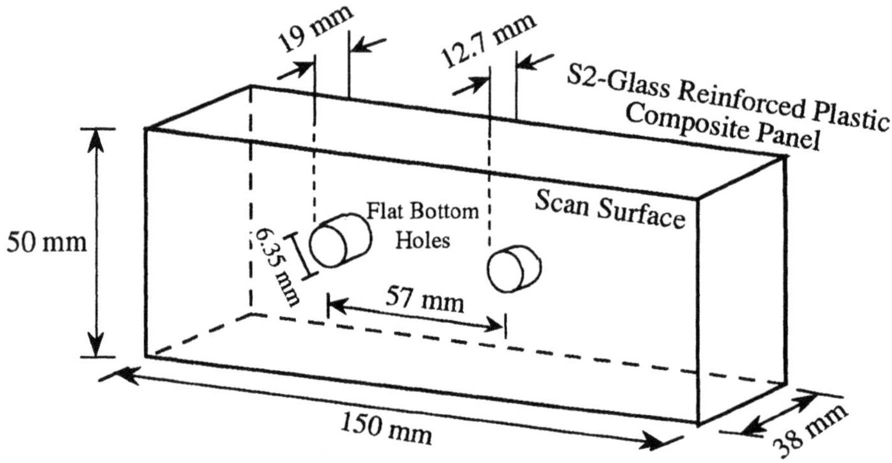

Fig. 5.13 Schematic geometry of a thick composite panel with two flat bottom holes (Qaddoumi *et al.*, 1995b).

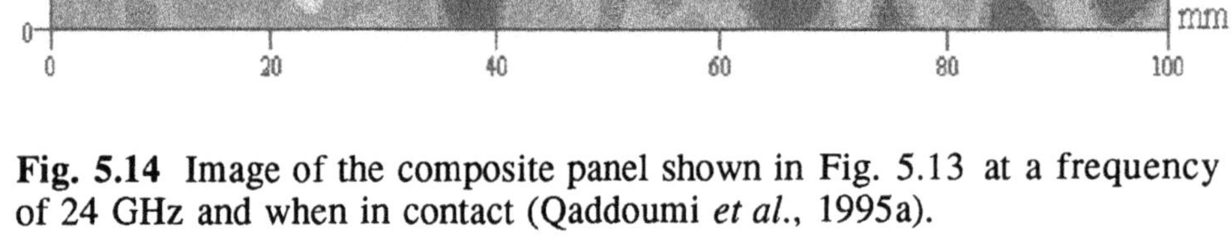

Fig. 5.14 Image of the composite panel shown in Fig. 5.13 at a frequency of 24 GHz and when in contact (Qaddoumi *et al.*, 1995a).

sandwich composites is a disbond in between any given two layers or delamination in a core layer. Inspection of thick composites, using near-field microwave techniques employing open-ended rectangular waveguide probes has received much attention during the past few years (Ganchev *et al.*, 1995b; Qaddoumi, Zoughi and Carriveau, 1996; Brown *et al.*, 1998; Easter *et al.*, 1998; Brown *et al.*, 1997; Zoughi, Lebowitz and Lukes, 1995; Lebowitz *et al.*, 1996; Qaddoumi *et al.*, 1998; Qaddoumi *et al.*, 1999). In many of these investigations near-field images of defective regions within various thick sandwich composites have been produced.

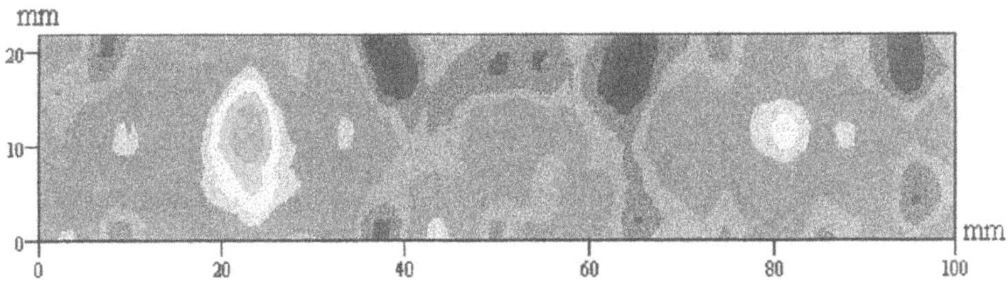

Fig. 5.15 Image of the composite panel shown in Fig. 5.13 at a frequency of 24 GHz and at a standoff distance of 0.5 mm (Qaddoumi *et al.*, 1995a).

The results have indicated the high potential of near-field, non-contact and one-sided microwave inspection techniques for detecting various embedded and real flaws including defect sizing capabilities.

As part of one of the earlier investigations in this area, several sandwich composite specimens were produced in which various flaws, representing disbonds and delaminations, were intentionally embedded. One of these specimens was made of a foam core of approximately 45 mm thick with a fiberglass epoxy laminate skin all backed by a conducting sheet. A circular 76.2 mm-diameter piece of 0.127 mm-thick mylar sheet was placed between the skin and the core to simulate a disbond at this interface. The edges of the circular mylar sheet were held to the core and the skin with a mylar tape of approximately 0.076 mm in thickness. This resulted in the edges of the mylar sheet to be slightly thicker than the rest of it (Ganchev *et al.*, 1995b).

Many experiments were conducted at various frequencies and standoff distances to detect this flaw and make an image of it. Figure 5.16 shows the image of this disbond at a frequency of 24 GHz. Due to the surface variations in this region of the specimen, a 1 mm-thick plexiglass sheet was used to keep a constant standoff distance. The scanning area was 154 mm by 150 mm. The disbonded region is clearly shown in this image. The edges of the circular mylar sheet are also seen to be different than the rest of the sheet, as expected due to the mylar tapes. The vertical feature outside of the circular disbonded region correspond to the fiberglass bundles in the skin. Notice that the electric field vector (i.e. polarization) coincided with the vertically shown bundles. The horizontal bundles are not seen in this image since they are orthogonal to the electric field vector associated with the waveguide aperture probe. It is also important to note that the size of the circular disbond provided by the image corresponds

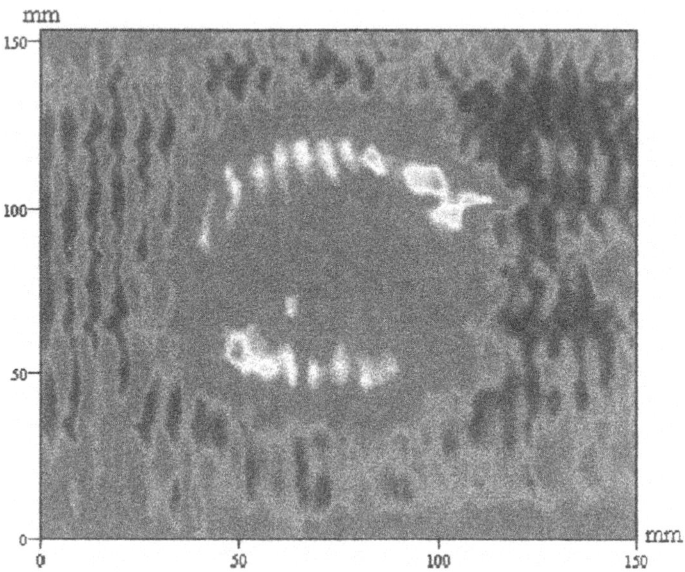

Fig. 5.16 Image of the disbond simulated by a circular mylar sheet in a sandwich composite (Ganchev *et al.*, 1995b).

very well with the actual size of the mylar sheet diameter. The level of detail provided in this image is a strong indication of the ability and potential of microwave imaging techniques for detection of flaws in composite structures.

5.4.4 Impact damage in thick sandwich composite

The above composite specimen was also used to investigate the potential of detecting impact damage in sandwich composites. The surface impact damage was created, after cure, by dropping a 5 kg steel ball on the sample with a 67.8 J energy (Ganchev, Runser, Qaddoumi, *et al.*, 1995b). Figure 5.17 shows the image of the impacted region at a frequency of 24 GHz using a 3.2 mm-thick plexiglass sheet to provide for a constant standoff distance. The results clearly show the impacted area and its boundaries. The signals from the warp/weave crossovers in the fiberglass epoxy laminate skin, indicate the damage at these locations. This may be a result of induced debond and/or porosity at these locations. Similar phenomenon has also been assumed in another study in which cyclical impact was induced in a composite specimen (Radford *et al.*, 1994). Figure 5.18

shows the intensity plot (3-D surface plot) of this impacted region. A spherical indentation in the center can be clearly seen. This is assumed to be as a result of the spherical shape of the steel ball which was used to induce the impact damage.

These results clearly demonstrate the potential of near-field microwave inspection techniques along with simple and quick image production (i.e. C-scan data) to detect, evaluate and analyze the properties of various embedded defects in thick sandwich composite structures.

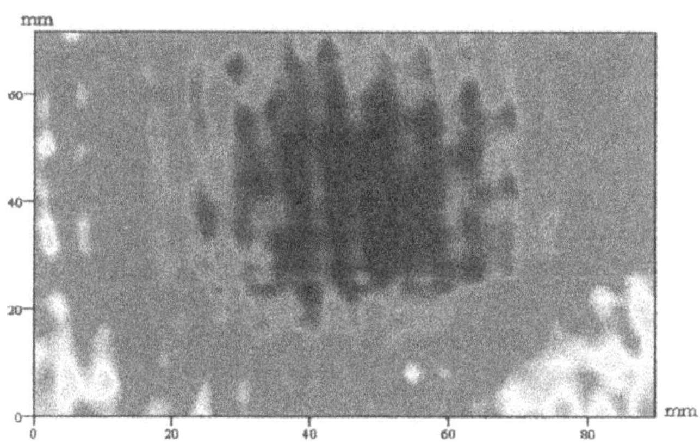

Fig. 5.17 Image of the impact damage in a sandwich composite (Ganchev *et al.*, 1995b).

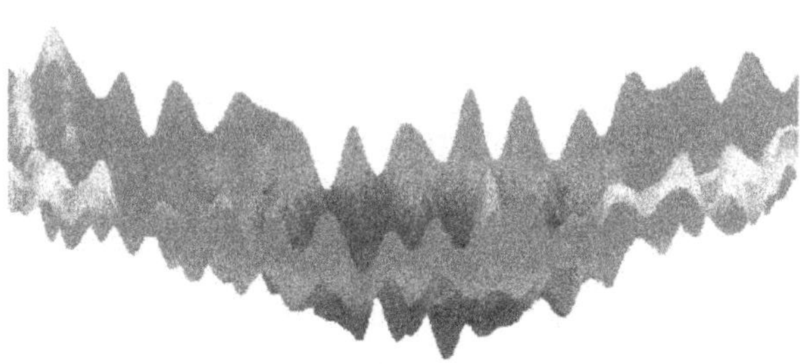

Fig. 5.18 Intensity plot (3-D surface plot) of the impact damage in a sandwich composite (Ganchev *et al.*, 1995b).

5.4.5 Localized porosity[4]

In Chapter 2 it was demonstrated that microwave dielectric characterization techniques can be used to determine distributed porosity in polymer composites (Gray *et al.*, 1995). Near-field microwave imaging can also be used to qualitatively distinguish among various localized porosity levels in polymer composites since in most practical cases the porosity is clustered (local) particularly in regions where fatigue and stress damage may have occurred. In this case the sensitivity of detection depends (besides the volume percentage of air) on the overall volume of the porous inclusion and its location (depth) within the composite. Consequently, an epoxy resin disk with a diameter of 76.5 mm and a thickness of 8.2 mm was produced with three porous inclusions embedded in it. The inclusions were in the shape of a pill with a diameter of 6.35 mm and thickness of 4.45 mm. These inclusions were made of air-filled microballoons providing three clustered porosity levels of 44%, 49% and 56%, as shown in Fig. 5.19. The distance between the centers of each inclusion was 19 mm, and in the thickness direction they were all located in the middle of the disk. An area of 56 mm by 18 mm was scanned at 35 GHz using an open-ended waveguide sensor. This frequency, in the 26.5-40 GHz (Ka-band) range, was chosen since for this range the waveguide aperture is 7.1 mm by 3.5 mm and provides for a higher spatial resolution compared with those at lower frequencies. An image of this specimen was produced, as shown in Fig. 5.20. The three inclusions are clearly seen. The colors (intensity) associated with each inclusion gives a qualitative measure of the difference in the porosity levels among the three inclusions. The image also shows the relative dimension of the inclusions to be very close to their actual sizes. Also, the distance between the inclusions in the image and the actual distance between them in the specimen correspond well to each other. Figure 5.21 shows the intensity plot which clearly shows the three inclusions as well as providing relative and quantitative information about the differences among them. The results effectively show the potential of using microwave non-destructive techniques not only for detecting local porosity, but also for quantitative estimate of the air content associate with it. Such an image, once calibrated may indicate absolute porosity level, and the size of the porous inclusion.

The results of this investigation, in addition to those presented in Chapter 2 for porosity evaluation, indicate the capabilities of microwave imaging as well as microwave material characterization techniques for evaluating porosity in polymers, and other similar dielectric materials.

---

[4] Portions of this sub-section are reprinted with permission from *Materials Evaluation*, vol. 53, no. 3, pp. 404-08, Copyright 1995 © The American Society for Nondestructive Testing, Inc.

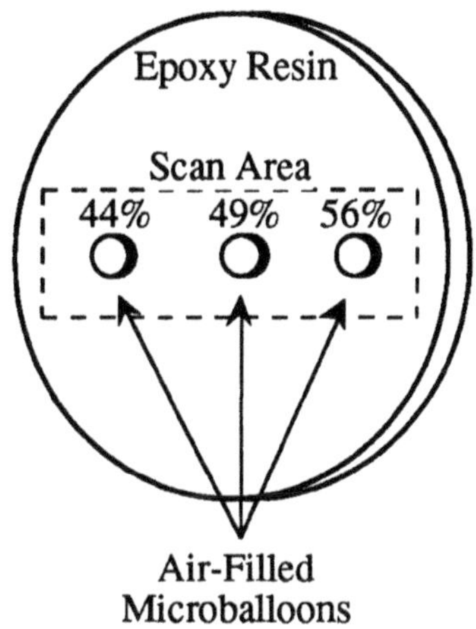

Fig. 5.19 The schematic of the epoxy resin specimen with three air-filled microballoon inclusions containing different levels of porosity (Gray *et al.*, 1995).

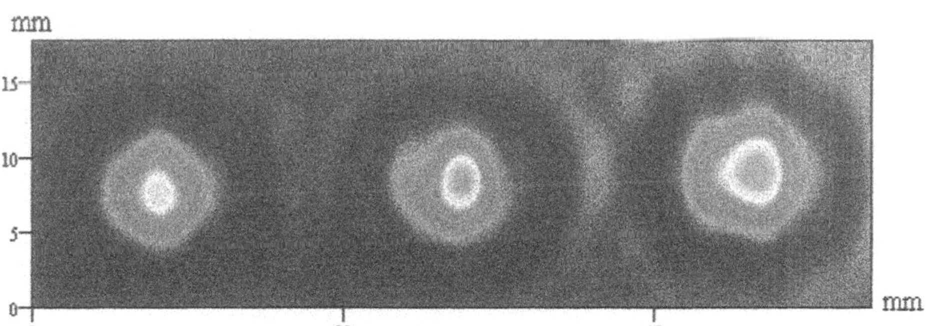

Fig. 5.20 The image of the specimen shown in Fig. 5.19 at a frequency of 35 GHz and in an in-contact fashion (Gray *et al.*, 1995).

Fig. 5.21 Intensity plot of the image shown in Fig. 5.20.

5.4.6 Resin variation in low-density fiberglass composites[5]

In Chapter 2 it was shown that microwave dielectric characterization techniques can be used to determine resin level in low-permittivity, low-loss and low-density fiberglass composites (Qaddoumi, Ganchev and Zoughi, 1996). To experimentally demonstrate the utility of near-field microwave imaging (using an open-ended rectangular waveguide aperture sensor) to achieve the same, a 25.4 mm-thick conductor backed fiberglass sample with 18.6% resin binder (i.e. base material) was prepared. This type of non-contact and one-sided measurement/imaging approach is well suited for on-line manufacturing and process control applications. Four regions of this sample were first removed, and subsequently were embedded with fiberglass with different resin binder levels, as shown in Fig. 5.22. These inclusions had square shapes with sides of approximately 25.4 mm and the distance between them was 50.8 mm. Since this was a real fiberglass product/sample, the density of the fiberglass in theses regions were less than the compacted ones used for dielectric property measurements (see Chapter 2). This means there was more air in these samples which further contributes to a reduction in their dielectric property values and the difference among them. Thus, the successful use of standoff distance analysis will be more indicative of its usefulness for distinguishing among similar low-permittivity, low-loss and low-density dielectric materials.

Subsequent to conducting a standoff distance analysis a microwave image of this sample was produced at a frequency of 24 GHz and at a standoff distance of approximately 4 mm, as shown in Fig. 5.23.

---

[5] Portions of this section are reproduced with permission from (Research in Nondestructive Evaluation, Microwave Diagnosis of Low Density Glass Fibers with Resin Binder, Qaddoumi, Ganchev and Zoughi, vol. 8, no. 3, pp. 177-88) © 1996 Springer-Verlag.

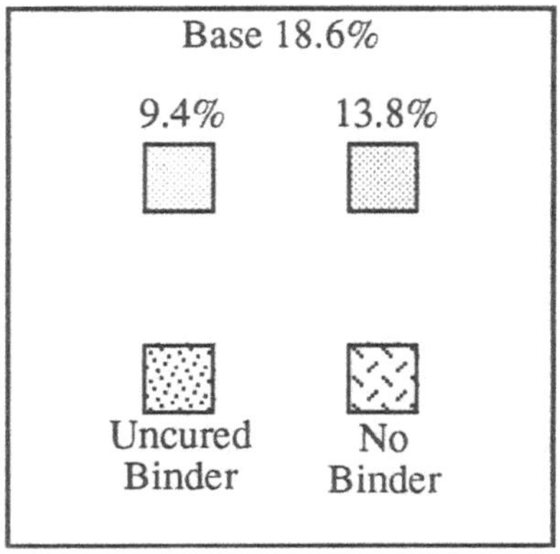

Fig. 5.22 Fiberglass specimen with four inclusions used for the non-contact and one-sided microwave imaging investigation (Qaddoumi, Ganchev and Zoughi, 1996).

The different colors in the image are proportional to the dielectric properties of each region, and hence related to the difference in their respective resin binder levels. This simple image illustrates the potential of using such a non-destructive and non-contact testing method for distinguishing among different resin binder contents in a 25.4 mm-thick low-density conductor backed fiberglass sheet for on-line manufacturing and process control applications. Figure 5.24 shows another image of this sample at a standoff distance of 10 mm and a frequency of 10 GHz. The voltage readings that produced these images may be calibrated to obtain an estimate of the resin binder level associated with each inclusion regions. The results indicate a very important point. Even though some of the regions in this specimen have close dielectric properties, using a non-contact approach can accentuate their differences and result in images that distinguish the different fiberglass specimens very effectively.

5.4.7 Rust under paint[6]

Accurate, reliable and early detection of rust/corrosion under paint and laminate composites, in many industrial and military applications, is of great concern and interest. To demonstrate the utility of near-field microwave non-destructive techniques, using open-ended rectangular waveguide sensors, a rusted steel specimen was used while painted over by

[6] Portions of this section are reproduced with permission from (Research in Nondestructive Evaluation, Microwave Detection of Rust Under Paint and Composite Laminates, Qaddoumi, Shroyer and Zoughi, vol. 9, no. 4, pp. 201-12) © 1997 Springer-Verlag.

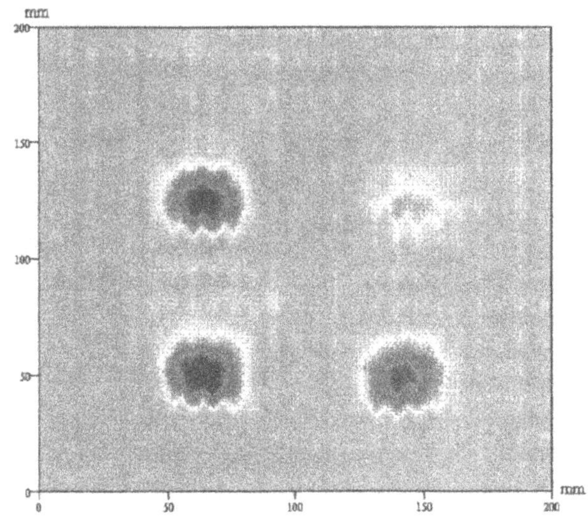

Fig. 5.23 Image of the fiberglass specimen in Fig. 5.22 at a standoff distance of 4 mm and a frequency of 24 GHz (Qaddoumi, Ganchev and Zoughi, 1996).

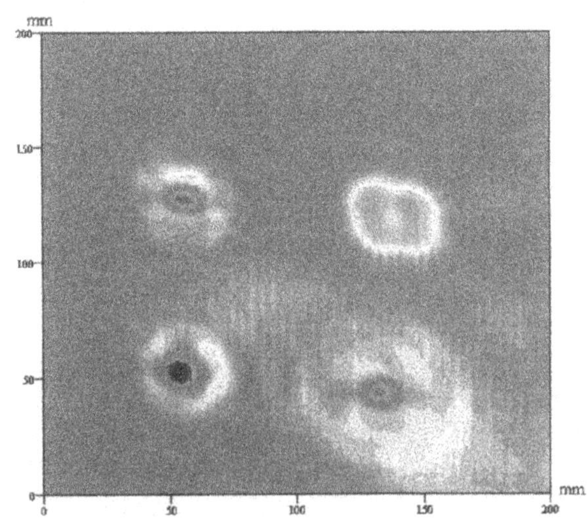

Fig. 5.24 Image of the fiberglass specimen in Fig. 5.22 at a standoff distance of 10 mm and a frequency of 10 GHz (Qaddoumi, Ganchev and Zoughi, 1996).

common spray paint to as much as ten coats of paint. These experiments were conducted at 24 GHz and 10 GHz (Qaddoumi, Shroyer and Zoughi, 1997).

Figure 5.25 shows a steel specimen with a 40 mm by 40 mm area of rust. This specimen was produced by acquiring a relatively flat piece of steel on which a thin layer of rust had already been produced (naturally). The 40 mm by 40 mm area was then masked out by a piece of tape and the remaining surface was sand blasted. The average thickness of the rust layer was measured (using a micrometer) to be approximately 0.08 mm. Subsequently, this specimen was painted with up to 0.60 mm (representing ten painting cycles) of common spray paint, as uniformly as possible. After applying various layers of paint microwave images of the rust specimen were produced using raster scans (every 2 mm by 2 mm) of the specimen at 24 GHz and 10 GHz. These measurement systems were designed with an open-ended waveguide aperture as their sensors while producing a dc voltage primarily proportional to the phase of the reflection coefficient measured at the waveguide aperture.

Figure 5.26 shows the image of this specimen at a standoff distance of 4 mm and at 24 GHz when coated with a paint thickness of 0.292. The rusted/corroded area is clearly visible in the center of the image. There is an elongated region in the upper right hand corner of the image which shows up as a region with a color in-between paint and rust. The steel specimen had a very subtle indentation in this region. Consequently, one can consider this region of having a slightly thicker paint than the rest of the painted areas. Therefore, this image not only shows the clear possibility of detecting a thin layer of rust under paint, they also illustrate the fact that paint thickness variation can be distinguished from the presence of rust. This has been shown to be possible using the multi-layered composite analysis presented in Chapter 3 (Qaddoumi, Shroyer and Zoughi, 1997). Figure 5.27 shows a similar image of the specimen with a paint thickness of 0.267 mm at 24 GHz and a standoff distance of 12.5 mm, indicating the ability to perform these measurements at higher than a few mm standoff distances.

Figure 5.28 shows a scan at a frequency of 10 GHz and a standoff distance of 2 mm for a paint thickness of 0.292 mm. Once again the rusted area is clearly visible. It should be noted that the rusted area looks relatively smaller than those at 24 GHz. This is as a direct result of the ratio of the dimensions of the rusted area to those of the waveguide aperture at this frequency.

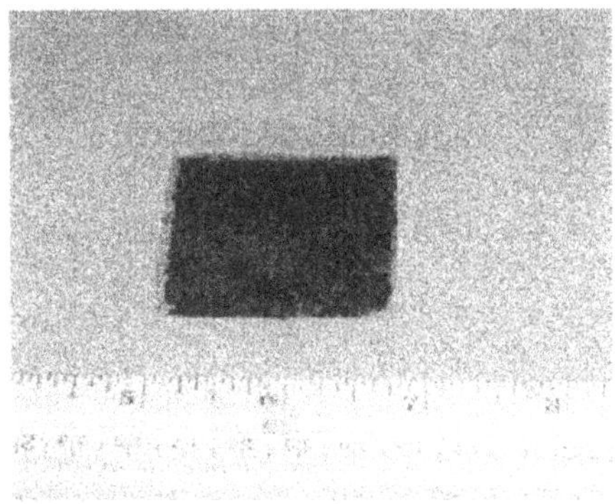

Fig. 5.25 A 40 mm by 40 mm area of rust on a steel plate (Qaddoumi, Shroyer and Zoughi, 1997).

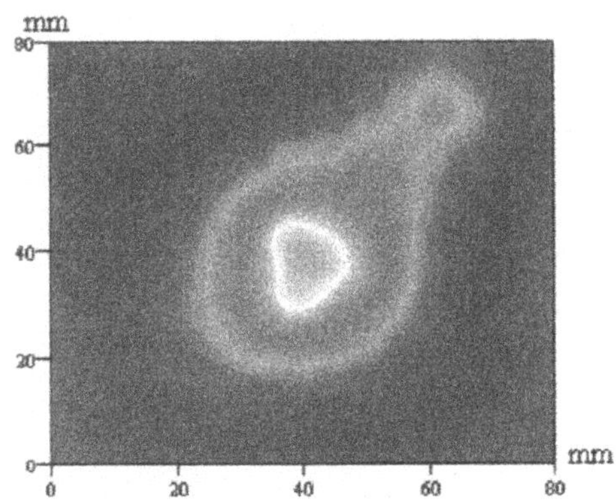

Fig. 5.26 Image of the rust shown in Fig. 5.25 under 0.292 mm of paint, at a frequency of 24 GHz and a standoff distance of 4 mm (Qaddoumi, Shroyer and Zoughi, 1997).

5.4.8 Rust under laminate composites

In some applications it is also important to detect rust under relatively thick composite laminates. The applicability of microwave techniques was investigated for this purpose as well. Another patch of rust similar in dimensions to the one mentioned in the previous section was placed under a low-permittivity synthetic rubber composite with a thickness of 25.4 mm. Using the same measurement setup an image of the rust specimen was obtained at a frequency of 10 GHz and at a standoff distance of 2 mm, as

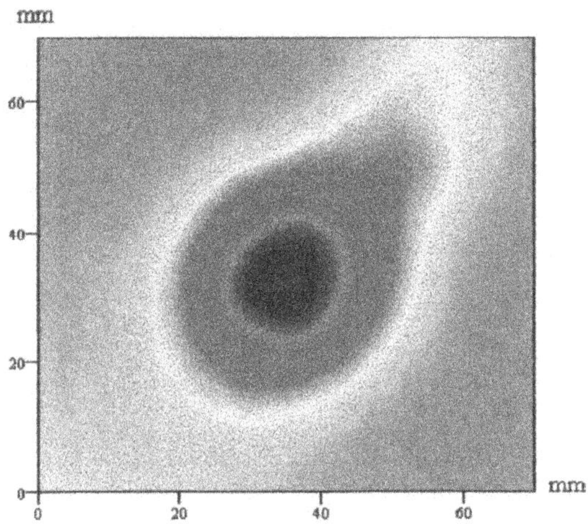

Fig. 5.27 Image of the rust shown in Fig. 5.25 under 0.267 mm of paint, at a frequency of 24 GHz and a standoff distance of 12.5 mm (Qaddoumi, Shroyer and Zoughi, 1997).

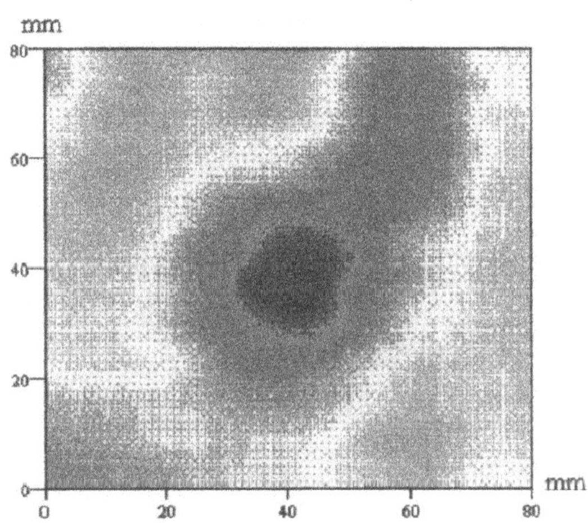

Fig. 5.28 Image of the rust shown in Fig. 5.25 under 0.292 mm of paint, at a frequency of 10 GHz and a standoff distance of 2 mm (Qaddoumi, Shroyer and Zoughi, 1997).

shown in Fig. 5.29. In addition, an image at a frequency of 24 GHz and at a standoff distance of 6 mm was also produced, as shown in Figure 5.30. In both cases the presence of rust is clearly detected along with good information about the extent of the rust patch.

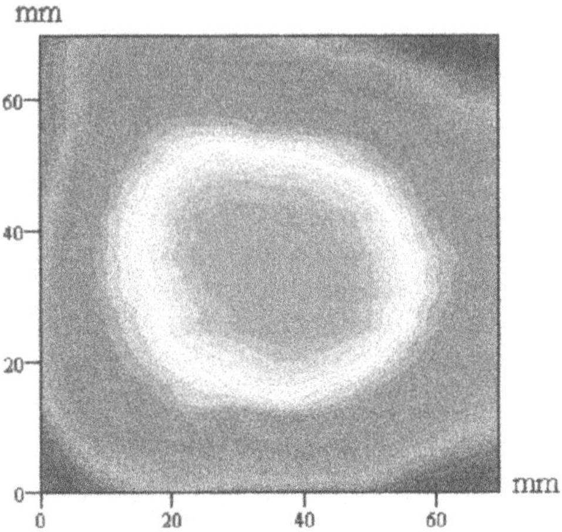

Fig. 5.29 Image of the rust under 25.4 mm of a low-permittivity synthetic rubber composite, at a frequency of 10 GHz and a standoff distance of 2 mm.

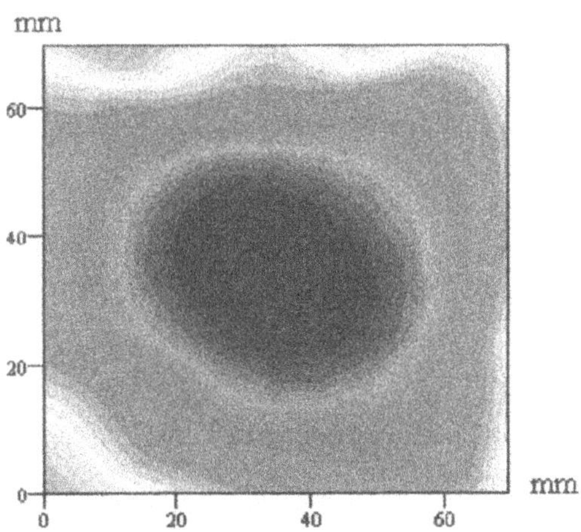

Fig. 5.30 Image of the rust under 25.4 mm of a low-permittivity synthetic rubber composite, at a frequency of 24 GHz and a standoff distance of 6 mm.

5.5 ISSUES ASSOCIATED WITH NEAR-FIELD MEASUREMENTS AND IMAGING

In this chapter several illustrative examples were provided which indicate the capability of near-field microwave imaging for non-destructive, non-contact and one-sided evaluation of various materials. In all cases, when producing false color (contrast) images (i.e. raster scan) of these materials high image contrast as well as high spatial resolution were obtained. These images provided direct information about the size of an inclusion and relative distance between any two given anomalies. It was also observed that thin flaws such as disbonds and rust under paint are easily and effectively detected. The thickness of these anomalies is in the direction of propagation. Thus, it is shown that even though when the ratio of the anomaly thickness to the operating wavelength is very small it is robustly detected and many subtle features of the anomaly are also revealed. This is as a direct result of operating in the near-field of a probe, such as an open-ended rectangular waveguide, in which the field properties (i.e. phase and magnitude) are strongly dependent on the distance away from the probe. This important characteristic was effectively demonstrated throughout Chapter 3.

When producing raster scans (referred to as false color contrast images in this chapter), using open-ended rectangular waveguides probes, one must be cognizant of a few important issues and practical considerations. The following is a list of some of the more important issues:

- When operating in the near-field region of open-ended probes, the spatial resolution is determined by the probe dimensions as opposed to the frequency (or wavelength) of operation. It may be argued that probe dimensions are dependent on the operating frequency. This is not universally true about all probes. For example, the broad dimension of a rectangular waveguide is a function of its dominant mode cutoff frequency. The narrow dimension of the guide does not play an important role, as it relates to launching dominant mode signals (Pozar, 1998). Coaxial probes, on the other hand, can operate in relatively much broader operating frequencies for a given probe dimension. When using tapered waveguides or small resonant apertures, very high spatial resolutions can be obtained (Golosovsky, Galkin and Davidov, 1996; Ash and Nicholls, 1972; Tabib-Azar, Shoemaker and Harris, 1993; Bryant and Gunn, 1965).
- The probe aperture is usually exposed to a certain area of the specimen under evaluation. Therefore, the aperture acts as a spatial integrator over the exposed area. Consequently, the signal received from any such region of the specimen is integrated by this aperture. This feature also allows conducting measurements with a relatively coarse scanning steps (i.e. faster image production) without a significant loss in the information revealed by the image.
- The electric field distribution, at the aperture of the probe also plays an important role in these measurements. This distribution influences the spatial resolution, and the manner by which the aperture integrates over

the testing region. For example, the half sine wave distribution over a rectangular waveguide aperture non-linearly integrates over the testing region (Qaddoumi and Zoughi, 1997).

- When considering the field properties inside a material, it has been shown that for open-ended rectangular waveguide probes, the electric field remains confined within the aperture dimensions for a good distance into the material (Qaddoumi *et al.* 1995b). In addition, the field remains confined within the aperture dimensions for larger distances away from the aperture (i.e. within the interior of a material) as a function of increasing permittivity of the material under test (i.e. ceramics as opposed to plastics). This is an important issue since for most practical applications an interior anomaly is usually near the surface (i.e. disbond), and consequently high spatial resolutions are also achieved when detecting sub-surface anomalies. It has also been shown that probe radiation characteristics possess side-lobes which may interact with interior anomalies and produce extra features in an image (Qaddoumi *et al.* 1995b). Furthermore, these side-lobes may be present in one direction and not in the other, as shown in Figs. 5.9 and 5.10. This fact is also dependent upon whether significant higher-order modes are generated at the probe aperture. The presence of these modes may produce side-lobes in directions that the dominant mode field properties may not produce. The types, numbers and field properties of these higher-order modes depend on the frequency of operation (e.g. in a waveguide band) and the dielectric properties of the material under test (Bois, Benally and Zoughi, 1998).
- It is also important to consider the relative dielectric and conducting properties of an anomaly within a material relative to those of the material. Flaws and anomalies that are highly conducting and possess defined edges produce significant edge scattering which when combined with the probe radiation pattern can produce features in an image that may be unexpected and may be misinterpreted.
- As mentioned earlier, open-ended rectangular waveguide probes, have aperture dimensions that are frequency dependent. Thus, it is reasonable to assume that at higher frequencies, where the aperture dimensions become smaller, the spatial resolution becomes finer as well. This is indeed true based on all that has been mentioned thus far. However, form the practical point of view there are several issues that must be considered when higher frequencies are used. First, for most practical applications the increase in the resolution may not be as significant as one may expect when operating at higher frequencies. Second, the hardware cost is usually higher at higher frequencies (and not linearly as a function of frequency either). Hardware and components are less readily available at higher frequencies. Some

components may actually have to be special ordered. Third, when incorporating a standoff distance in a given measurement, at higher frequencies less variation in this parameter is usually tolerated than at lower frequencies. There are several techniques to keep standoff distance fairly constant throughout a scan. One approach involves the use of a mechanical (i.e. roller) system that holds the distance between the probe and the material under test constant. The roller system is not desirable when the material under test possesses relatively significant surface roughness (Qaddoumi, Bigelow, Mirshahi *et al.*, 1998). Another approach involves the use of extra circuitry that electronically compensates for the changes in standoff distance as a scan is performed (Qaddoumi, Bigelow, Frank *et al.*, 1999).

- In some special applications, one may attach a thin sheet of a dielectric to the probe aperture. This addition can be optimally made (i.e. optimum thickness and dielectric properties of the dielectric sheet) such that measurements can become more sensitive to detecting certain hard to detect anomalies such as very thin disbonds. This is analogous to biasing the system at a very sensitive, to the presence of a flaw, operating point. The use of Teflon tape, at the aperture, can also accommodate the requirement of keeping the standoff distance constant wile scanning a specimen.
- Raster scan type of image production is not the only approach to near-field imaging. One may also produce similar images using 1-D or 2-D array of sensors. This can also be achieved using scanning (phased) arrays (Bolomey and Pichot, 1990). The use of any of these techniques will depend upon many practical factors such as image production speed, how much system complexity can be tolerated, cost, etc.
- The inverse problem associated with near-field imaging is also an important issue worthy of a brief discussion. The inverse problem refers to extracting spatial and other information about an anomaly within a material from its image. In the far-field approach the electric field properties do not change as a function of directions transverse to the direction of propagation, and in the propagation direction the antenna pattern remains unchanged. Therefore, in this case one may readily obtain the radiator point spread function and use this information in a de-convolution of the image and this point spread function to obtain the exact spatial information about the material or the target that was imaged. On the other hand, and as mentioned earlier, the electric field properties radiated by a sensor in its near-field change as a function of all three spatial dimensions. Therefore, to perform the same de-convolution, the probe point spread function information must be known or obtained for every possible distance into

the material being imaged. Since one does not know where inside a material a defect may exist, this approach is rendered inefficient (or practically useless). Recently, an effort has been initiated to overcome the difficulties associated with this inverse problem (Mohammad-Djafari, Qaddoumi and Zoughi, 1999).

It is clear from the above discussions that near-field imaging, although a very powerful non-destructive tool, its foundations must be well understood so that images can be interpreted correctly. More work in this area is ongoing and further improvements to this technique are expected in the near future.

REFERENCES

Ash, A.E. and Nicholls, G. (1972) Super-resolution aperture scanning microscope. *Nature*, **237**, pp. 510-12.

Baker-Jarvis, J. (1990) Transmission/reflection and short-circuit line permittivity measurements, NIST Technical Note 1341, U.S. Government Printing Office, Washington, D.C., July.

Bakhtiari, S., Ganchev, S.I. and Zoughi, R. (1993) Microwave swept-frequency optimization for accurate thickness or dielectric property monitoring of conductor-backed composites. Materials Evaluation, **51** (6), pp. 740-43.

Bois, K.J., Benally, A.D. and Zoughi, R. (1998) A multi-mode solution for the analysis of the reflection coefficient of open-ended rectangular waveguides radiating into a dielectric infinite half-space. *Proceedings of the Progress in Quantitative Nondestructive Evaluation*, **17A**, pp. 705-11.

Bolomey, J.C. and Pichot, C. (1990) Microwave tomography: from theory to practical imaging systems. *International Journal of Imaging Systems and Technology*, **2**, pp. 144-56.

Brown, L.M., DeLoach, J.J., Qaddoumi, N. *et al.* (1998) An overview of microwave inspection of Naval multi-layered composites. *Proceedings of the Third Pacific Rim International Conference on Advanced Materials and Processing*, pp. 1667-72, Hawaii.

Brown, L.M., DeLoach, J.J., Zoughi, R. *et al.* (1997) Preliminary evaluation of microwave techniques for inspection of thick layered composite deck joint. *Proceedings of the Eighth International Symposium on Nondestructive Characterization of Materials*, Vol. III, pp. 305-10, Boulder, CO.

Bryant, C.A. and Gunn, J.B. (1965) Noncontact technique for local measurement of semiconductor resistivity. *Review of Scientific Instruments*, **36**, pp. 1614-17.

Carriveau, G.W. (1994) Benchmarking of the state-of-the-art in nondestructive testing/evaluation for applicability to in the composite armored vehicle (CAV) advanced technology demonstrator (ATD) program. NTIAC Report 7304-104:GWC-D172.17.

d'Ambrosio, G., Massa, R., Migliore, M.D. *et al.* (1993) Microwave defect detection on low-loss composites. *Materials Evaluation*, **51** (2), pp. 285-89.

Easter, J., Qaddoumi, N., Zoughi, R. *et al.* (1998) Preliminary results of detecting and locating defects under triangular thick composite structural members. *Proceedings of the SPIE Symposium on Nondestructive Evaluation Techniques for Aging Infrastructure Manufacturing*, **3396**, pp. 129-34, San Antonio, TX.

Ganchev, S.I., Qaddoumi, N., Bakhtiari, S. *et al.* (1995a) Calibration and measurement of dielectric properties of finite thickness composite sheets with open-ended coaxial sensors. *IEEE Transactions on Instrumentation and Measurement*, **44** (6), pp. 1023-29.

Ganchev, S.I., Runser, R.J., Qaddoumi, N. *et al.* (1995b) Microwave nondestructive evaluation of thick sandwich composites. *Materials Evaluation*, **53** (4), pp. 463-67.

Gardiol, F.E. (1984) Introduction to microwaves, Artech House, Dedham, MA.

Golosovsky, M., Galkin, A. and Davidov, D. (1996) High-resolution resistivity mapping of large-area YBCO films by a near-field millimeter-wave microscope. *IEEE Transactions on Microwave Theory and Techniques*, **44** (7), pp. 1390-92.

Gopalsami, N., Bakhtiari, S., Dieckman, A. *et al.* (1994) Millimeter-wave imaging for nondestructive evaluation of materials. *Materials Evaluation*, **52** (3), pp. 412-15.

Gray, S, Ganchev, S.I., Qaddoumi, N. *et al.* (1995) Porosity level estimation in polymer composites using microwave. *Materials Evaluation*, **53** (3), pp. 404-08.

Jenkins, S., Warham, A. and Clarke, R. (1992) Use of open-ended coaxial line sensor with a laminar or liquid dielectric backed by a conducting plane. *Proceedings of Institute of Electrical Engineers*, **139** (2), pp. 179-82.

King, R.J. and Yen Y.H. (1981) Probing amplitude, phase, and polarization of microwave field distributions in real time. *IEEE Transactions on Microwave Theory and Techniques*, **29** (11), pp. 1225-30.

Kraszewski, A., Stuchly, M. and Stuchly, S. (1983) ANA calibration method for measurements of dielectric properties. *IEEE Transactions in Instrumentation and Measurement*, **32**, pp. 385-87.

Lebowitz, C., Zoughi, R., Spicer, J.W.M. *et al.* (1996) Comparison of ultrasonic, microwave and photothermal imaging of defective graphite-epoxy composite panels. *Proceedings of SPIE Symposium, Nondestructive Evaluation Techniques for Aging Infrastructure and Manufacturing Conferences*. **2944**, pp. 67-74, Scottsdale, AZ.

Liping, L., Deming, X. and Zhiyan, J. (1986) Improvement in dielectric measurement technique of open-ended coaxial line resonators method. *Electronic Letters*, **22** (7), pp. 373-75.

Mohammad-Djafari, A., Qaddoumi, N. and Zoughi, R. (1999) Deconvolution approach for resolution enhancement of near-field microwave images. *Proceedings of the SPIE, Mathematical Modeling, Bayesian Estimation, and Inverse Problems Conference*, **3816**, Denver, CO.

Mosig, J., Besson, J.C., Gex-Farby, M. *et al.* (1981) Reflection of an open-ended coaxial line and application to nondestructive measurement of materials. *IEEE Transactions on Instrumentation and Measurement*, **30**, pp. 46-51.

Nyshadham, A., Sibbald, C. and Stuchly, S. (1992) Permittivity measurements using open-ended sensors and reference liquid calibration - An uncertainty analysis. *IEEE Transactions on Microwave Theory and Techniques*, **40** (2), pp. 305-13.

Pozar, D.M. (1998) Microwave engineering, John Wiley and Sons, New York, NY.

Qaddoumi, N., Abiri, H., Ganchev, S. *et al.* (1995b) Near-field analysis of rectangular waveguide probes used for imaging. *Proceedings of the Review of Progress in Quantitative NDE*, **15A**, pp. 727-32, Plenum Press, NY.

Qaddoumi, N., Bigelow, T., Mirshahi, R. *et al.* (1998) Microwave inspection of thick composites and deck joints and development of a first generation microwave prototype inspection system. *Final Report*, Naval Surface Warfare Center, Carderock Division, Survivability, Structures and Materials Directorate, Bethesda, MD.

Qaddoumi, N., Bigelow, T., Frank, M. *et al.* (1999) Elimination of sensitivity to surface roughness variations in thick composite deck joints. *Final Report*, Naval Surface Warfare Center, Carderock Division, Survivability, Structures and Materials Directorate, Bethesda, MD.

Qaddoumi, N., Ganchev, S.I. and Zoughi, R. (1996) Microwave diagnosis of low-density fiberglass composites with resin binder. *Research in Nondestructive Evaluation*, **8** (3), pp. 177-88.

Qaddoumi, N., Ganchev, S.I., Carriveau, G.W. *et al.* (1995a) Microwave imaging of thick composite panels with defects. *Materials Evaluation*, **53** (8), pp. 926-29.

Qaddoumi, N., Shroyer, A. and Zoughi, R. (1997) Microwave detection of corrosion under paint and composite laminate coatings. *Research in Nondestructive Evaluation*, **9** (4), pp. 201-12.

Qaddoumi, N., Zoughi, R. and Carriveau, G.W. (1996) Microwave detection and depth determination of disbonds in low-permittivity and low-loss thick sandwich composites, *Research in Nondestructive Evaluation*, **8** (1), pp. 51-63.

Qaddoumi, N. and Zoughi, R. (1997) Preliminary study of the influences of effective dielectric constant and non-uniform probe aperture field distribution on near-field microwave images. *Materials Evaluation*, **55** (10), pp. 1169-73.

Radford, D., Barber, B., Ganchev, S.I. *et al.* (1994) Microwave nondestructive evaluation of glass fiber/epoxy composites subjected to impact fatigue. *Proceedings of SPIE International Symposium on Optics, Imaging and Instrumentation, Advanced Microwave and Millimeter Wave Detectors*, **2275**, pp. 21-26, San Diego, CA.

Stuchly, M. and Stuchly, S. (1980) Coaxial line reflection methods for measuring dielectric properties of biological substances at radio and microwave frequencies - A review. *IEEE Transactions on Instrumentation and Measurement*, **29**, pp. 176-83.

Tabib-Azar, M., Shoemaker, N.S. and Harris, S. (1993) Nondestructive characterization of materials by evanescent microwaves. *Measurement and Scientific Technology*, **4**, pp. 583-90.

Ulaby, F.T., Moore, R.K. and Fung, A.K. (1986) Microwave Remote sensing, passive and active. Volume 2, pp. 842-44, Artech House, Dedham, MA.

Weir, W.B. (1970) Automatic measurement of complex dielectric constant and permittivity at microwave frequencies, *Proceedings of IEEE*, **62** (11), pp. 33-36.

Zoughi, R., Cone, G.L. and Nowak, P.S. (1991) Microwave nondestructive detection of rebars in concrete slabs. *Materials Evaluation*, **49** (11), pp. 1385-88.

Zoughi, R., Lebowitz, C. and Lukes, S. (1995) Preliminary evaluation of the potential and limitations of microwave NDI methods for inspecting graphite composites. *Proceedings of the Seventh International Symposium on Nondestructive Characterization of Materials*. **210-213** (part 2), pp. 611-18, Prague, Czech Republic.

CHAPTER 6

Other developments and future

6.1 INTRODUCTION

Thus far in this text, four important issues of interest to the non-destructive testing community were discussed in detail; namely, a) dielectric material characterization, b) layered composite inspection, c) fatigue surface crack detection and evaluation in metals, and d) microwave measurement techniques and imaging. There have been other advances made in this field each worthy of a brief discussion. The interested reader is encouraged to consult with the respective references provided for each of these topics. This chapter is dedicated to these discussions along with some of the prevalent industrial and cultural issues associated with these techniques. Finally, a discussion (author's opinion) regarding the future of microwave non-destructive testing and evaluation techniques (in particular the near-field approach) is provided.

6.2 TESTING AND EVALUATION OF CEMENT BASED MATERIALS

One of the areas that has received considerable attention during the past few years has been the near-field microwave testing and evaluation of cement based materials. Several specific investigations were conducted which indicated the great promise and potential of using these techniques for inspecting cement based materials and structures. Zoughi, Cone and Nowak used a simple microwave non-destructive method, employing open-ended rectangular waveguide probes, to locate a reinforcing steel bar in a concrete slab (Zoughi, Cone and Nowak, 1991). They showed the ability to manipulate the polarization properties of the microwave signal to enhance the image of the steel bar. In addition, they detected the presence of a break in the steel bar. Their experiments also showed the sensitivity of microwave signals to aggregate size distribution in the slab. Subsequently, the potential of this microwave technique for detecting aggregate size distribution by manipulating the frequency of the microwave signal was demonstrated (Zoughi, Nast and Nowak, 1992). The influence of aggregate size distribution on the reflection coefficient decreases as a function decreasing frequency (i.e. larger wavelengths). A strong correlation between the magnitude of the reflection coefficient of the microwave signal and hardened cement paste water-to-cement (w/c) ratio was also shown. Subsequently, through extensive measurements using open-ended rectangular waveguides a correlation between the magnitude of reflection coefficient and the compressive strength of hardened cement paste (or w/c ratio) was obtained (Zoughi, Gray and Nowak, 1995). Later, the potential of using monopole antennas for the same purpose was demonstrated as well (Shalaby and Zoughi, 1995). Consequently, these promising results prompted a more extensive investigation in using near-field microwave nondestructive testing and evaluation techniques, employing open-ended rectangular waveguide probes (Zoughi *et al.*, 1998). The objective of this investigation was to determine the near-field microwave reflection coefficient (in an in-contact or non-contact fashion). In cement paste there is a desire to quantitatively evaluate the w/c ratio and ultimately the cure state and compressive strength of the hardened cement paste. In mortar the influence of adding sand to cement paste was investigated. Thus, determining w/c ratio, compressive strength, sand-to-cement (s/c) ratio and porosity have been the primary goals of studying mortar. Concrete may be considered homogeneous or heterogeneous depending on the frequency of operation and the aggregate size and volume distributions. In concrete, which was the ultimate goal of this study, the near-field microwave reflection properties by which w/c ratio, coarse aggregate-to-cement (ca/c ratio), aggregate size and volume distributions could be determined have been of interest. Other related investigations have included the detection of chloride added to the mixing water in mortar specimens. In addition, detection of grout in masonry blocks has also been conducted.

The following is a succinct description of the results of these investigations to-date:

Cement paste (water + cement powder)

- The magnitude of reflection coefficient increases as a function of decreasing w/c ratio for cured cement paste. At first glance this seems inconsistent with the fact that higher water content should render higher magnitude of reflection coefficient measured at a waveguide aperture. However, a closer look reveals that during the curing process water molecules bond with cement molecules, and the remaining free water evaporates. Thus, the water content becomes less and less free and more and more bound as a function of curing time. Free water has much higher dielectric properties compared to those of cement powder, whereas bound water has similar to cement powder dielectric properties. More of free water evaporates in higher w/c ratio specimens since there is less cement available to bind with. Thus, cured specimens with lower w/c ratio have more bound water and consequently higher reflection coefficient than those with higher w/c ratios.

- The magnitude of reflection coefficient has been found to be distinctly correlated to the w/c ratio of cement paste, and subsequently to its 28-day compressive strength (moist cured for 3 days in a hydration room and thereafter in air at room temperature) (Zoughi, Gray and Nowak, 1995).
- A simple expression predicting the microwave reflection properties of cement paste as a function of time has been obtained. Consequently, the w/c ratio of a cement paste specimen may be obtained by comparing two reflection coefficient measurements conducted several hours or a few days apart after the paste has been poured. In addition, it is possible to correlate the compressive strength of cement paste during curing to the measured microwave reflection properties (as a percentage of the 28-day strength) (Bois, 1999)

Mortar (water + sand + cement powder)

- A relationship between the standard deviation of the magnitude of reflection coefficient at higher frequencies and the s/c ratio of a mortar specimen, has been established. Information on the w/c ratio of mortar specimens is obtained when the average value of the measurements is taken at lower frequencies (Bois *et al.*, 1997).
- Mortar is a homogeneous dielectric mixture (even at 10 GHz). Thus, a simple dielectric mixing model has been obtained which predicts the constituent volume contents of a mortar specimen. Consequently, the porosity (volume content of distributed air) of a mortar specimen can be determined using this mixing model (Bois, Mirshahi and Zoughi, 1997).

Concrete (water + sand + coarse aggregate + cement powder)

- It has been determined that the probability distribution functions of the reflection properties of concrete, measured at high and low frequency bands, possess distinct and well-known distributions. At higher frequencies, the distribution is Gaussian whereas at low frequencies the distribution is uniform. With the use of the modifiable parameters in each of these distributions, the constituent volume distribution of a given concrete mixture can be obtained from its scattering characteristics (Bois, 1999; Bois, Benally and Zoughi, 1998).

- Similar to mortar, the results of the reflection property measurements indicate that the w/c ratio in concrete (hence its strength) can be correlated to the average value of the magnitude of reflection coefficient measured at several independent locations on a specimen at lower frequencies (i.e. 3 GHz). At lower frequencies the influence of aggregate size distribution is less on the measured magnitude of reflection coefficient than at higher frequencies since the aggregates electrically look smaller at lower frequencies.
- It has also been demonstrated that cure-state monitoring and fresh w/c ratio determination in concrete can be accomplished using this near-field technique (Bois *et al.*, 1998a). Using this approach one is potentially capable of determining the w/c ratio of a concrete batch a short time after the concrete has been poured. Cure-sate monitoring and fresh w/c ratio determination are two very important issues in the construction industry.
- Similar to mortar, the standard deviation and the statistical distribution of the measured magnitude of reflection coefficient at higher frequencies is a function of the aggregate size and volume distributions. Hence, the constituent volume fraction and distribution of a concrete specimen may be determined at higher frequencies (i.e. 10 GHz) (Bois, 1999; Zoughi *et al.*, 1998).
- A decision process, based on maximum likelihood approach, has been devised which correlates the statistical properties of concrete specimens at several frequencies to their material constituent makeups and respective compressive strengths (Bois, 1999, Zoughi *et al.*, 1998).
- Segregation of aggregates in concrete structures is an undesirable feature. Near-field microwave non-destructive techniques, employing open-ended rectangular waveguides, have shown great promise for detecting aggregate segregation as well (Bois *et al.*, 1999a). When aggregates segregate to the bottom portions of a vertical concrete structure, the statistics of the multiple measurements of the magnitude of reflection coefficient, at relatively high microwave frequencies such as 10 GHz, can be used to determine the degree of segregation. This method is fast and totally non-destructive and provides for a quick inspection of a large structure (as opposed to a few local measurements).

Chloride detection

- Chloride can be introduced into a concrete structure through various means including the mixing water, aggregates, cement, admixtures and the de-icing salts. In general, concrete provides adequate corrosion protection for steel reinforcing bars. However, in the presence of moisture and oxygen and when chloride is available the steel reinforcing bars corrode through an electrochemical process. Once the steel becomes corroded, the byproducts of corrosion will deteriorate the concrete. It is clearly of great interest to have a robust non-destructive technique to detect chloride in concrete as well as determine its penetration profile. Since microwave signals are also sensitive to the presence of salts in materials (such as water) it is expected that microwave non-destructive testing techniques can be used to detect its presence in concrete. To this end, preliminary experiments for detecting chloride contamination in mortar have been conducted for different mortar constituent makeups (Benally *et al.*, 1999). In these experiments, different amounts of salt were added to the mortar mixing water and measurements were conducted at 10 GHz and 3 GHz, respectively. The latter frequency seemed to provide for a linear correlation between the magnitude of reflection coefficient, measured using open-ended rectangular probes, and the level of salt (relative to the amount of cement). The results of these experiments clearly show the potential of these microwave techniques for this important inspection aspect of concrete.

Embedded sensors in cement based materials

- Embedded sensors such as modulated resonant dipoles can also be used in a variety of concrete testing and evaluation applications to provide valuable information about the state of a concrete structure. Modulated scattering techniques, employing resonant and near-resonant dipoles, have been effectively used in a variety of non-destructive applications (Bolomey and Pichot, 1990; Bolomey and Joachimowicz, 1994). Recently, an attempt has been made to use modulated scattering techniques for evaluating the dielectric properties of hardened cement paste (Joisel et al., 1999). In this experiment a near-resonant Pin diode-loaded dipole antenna was embedded about 2 cm away from the surface of a cement paste block as it was being prepared. This block was left in room temperature for twenty eight days. The dipole was then modulated at a slow rate of about 0.5 Hz and using an HP8510 vector network analyzer its reflection properties, measured by an open-ended rectangular waveguide probe at 7 GHz, were measured. It was shown that the dielectric properties (permittivity) of this specimen can be evaluated using this technique. This is a significant finding since this technique can be expanded to not only detect chloride in concrete but one can use a layered formulation as described in Chapter 3 to also evaluate the profile of chloride penetration. This work is ongoing at the time of writing of this text.

Masonry

- Masonry block cells are often required to be filled at certain prescribed intervals to provide a vertical structure with adequate lateral strength. In many cases these cells are also required to be reinforced with steel reinforcing bars. The cells are usually filled with grout (a high w/c ratio version of mortar). In many applications the cells may not be filled as prescribed which can cause weakness in the structure. In a recent investigation, it was effectively demonstrated that the same microwave technique as describes above can be used, at a frequency of near 3 GHz, to detect the presence or absence of grout in masonry blocks (Bois *et al.*, 1998). In addition, it was shown that a non-contact version of this approach can also detect the presence of grout. This is important when considering dry walls that may be covering a masonry wall.

Based on the promising results of these investigations one can use a simple, inexpensive, real-time and in-situ near-field microwave non-destructive technique to determine the in-situ compressive strength of a chosen region or location of a concrete structure (bridges, columns, pavement, beams, walls, etc.) as well as obtaining information about its constituent volumetric and distribution properties. Some of the characteristics and potential benefits of this near-field microwave technique are:

- these measurements are complete non-destructive, can be implemented at the site in real-time, and many measurements may be taken very quickly to increase the robustness of the results,
- many locations on a given test area may be examined very quickly,
- the entire testing system can be mounted on a small inspection cart or be backpack-carried while operated by a single individual with no specific microwave expertise,
- the testing system may be battery operated, thus expanding its portability and testing realm,
- the microwave hardware required to build such systems is relatively inexpensive,
- these methods give immediate results and thus alleviate testing turn around time and associated cost,
- the test may be done from a moving vehicle so that large areas such as roadways and bridges can be tested very quickly. This is due to the fact that in the case of an open-ended waveguide or a similar probe, no contact with the specimen is required (i.e. a standoff distance is present between the probe and the concrete specimen). Furthermore, the standoff distance value can be optimized to provide higher degree of measurement sensitivity (Bois *et al.*, 1999b),

- microwave signals (particularly at low frequencies) are insensitive to the presence of dust and smoke resulting in very little, to no test site preparation being necessary.
- the testing system can also become a permanent fixture within a structure (a waveguide or a monopole probe mounted inside a structure). This feature provides for periodic or continuous monitoring of the structure, and thus can become a sensor within a **smart** structure.

6.3 OPEN-ENDED COAXIAL PROBES FOR LAYERED COMPOSITE INSPECTION

As was mentioned several times in this text, open-ended rectangular waveguide probes are not exclusively used for near-field microwave non-destructive testing and evaluation. Another commonly used probe is an open-ended coaxial line probe. These probes offer certain advantages and possess certain limitations. Coaxial lines operate in a relatively large frequency range. The dimensions of a coaxial line set the limit on the highest frequency it may be operated at before higher-order modes are generated. The aperture of these probes, which is used for non-destructive testing purposes, is relatively small resulting in high spatial resolutions. In addition, coaxial lines are manufactured to be rigid, semi-rigid, semi-flex and flexible. Therefore, they can be made to reach hard-to-get to regions of a specimen under inspection. Open-ended coaxial probes are inefficient radiators (i.e. antennas), and unlike an open-ended rectangular waveguide the fields are concentrated near the probe aperture. These fields are considered to be quasi-static for most applications. In Chapter 2 many references pertaining to the utilization of this probe for dielectric material characterization were provided.

As it relates to non-destructive inspection of layered composites, these probes can play a significant role. However, when used for this purpose the layered composite can not be very thick or made of highly lossy materials. Therefore, these probes are suitable for determining accurate thickness of coatings (on top of conducting plates) and thickness evaluation of various layers in a thin multi-layered structure (Bakhtiari, Ganchev and Zoughi, 1994). One of the issues associated with open-ended coaxial lines is its calibration procedure. As mentioned in Chapter 5, this is not always a straightforward problem. A simple calibration approach in conjunction with layered composite inspection, has been developed which alleviates some of the traditional difficulties associated with open-ended coaxial probe calibration (Ganchev et al., 1995). It is anticipated that these probes will find their utility in applications where thin dielectric sheets and coatings such as paint or thermal barrier coating are used.

6.4 FATIGUE SURFACE CRACK DETECTION AND EVALUATION IN METALS USING OPEN-ENDED COAXIAL PROBES

Open-ended coaxial lines have also been successfully used for detecting and evaluating fatigue surface cracks in metals (Zoughi, Hayes and Ganchev, 1996). Many of the advantages of this probe that were mentioned in the previous section and those associated with microwave fatigue surface crack detection (as outlined in Chapter 4) make this an effective probe for this purpose. Recently, the interaction of such a probe with a metal surface in which a crack exists was modeled as well (Wang, 1999). In this modeling approach the crack is assumed to be a thin rectangular waveguide aperture exposed to the open-ended coaxial probe aperture. The length of the crack changes as it is scanned by the coaxial aperture. This causes the crack to continuously have a different cutoff frequency as it is being scanned. Mode matching technique and conservation of complex power at the junction between these two apertures are used to determine the reflection properties at the coaxial probe aperture resulting in the coax crack characteristic signal. The convergence of the numerical derivation is also studied. The results indicate the potential for crack detection optimization and crack sizing.

6.5 SURFACE INSPECTION OF ROLLED STEEL

Surface inspection of rolled steel is an important practical issue in steel industry. When producing steel sheets of various widths and thicknesses a wide variety of surface defects can be produced as well. These defects include holes, laminations, slivers, pock marks, pick ups, gouges, scratches, to name a few. Each of these defects has its own unique properties and dimensions. It is of great interest to develop a non-contact and on-line surface inspection technique which is capable of detecting and evaluating the properties of these defects. The manufacturing environment in which these rolled steel sheets are produced is not conducive to most non-destructive testing techniques. Depending on a particular point in the manufacturing process there may exist excessive heat, moisture and steam. In addition, the manufacturing process can be occurring at speeds in excess of 5000 ft/min. Therefore, one can appreciate the difficulties associated with this surface inspection. Non-contact, one-sided and on-line microwave testing and evaluation techniques have been used for this purpose. The preliminary results obtained from these investigations have shown tremendous potential for on-line, real-time and non-contact rolled steel surface inspection (Zoughi, *et al.*, 1997a; Zoughi *et al.*, 1997b). Many experiments were conducted on a specially manufactured test stand which allowed rolled steel sheet specimens to be attached to it and move at speeds of up to 1000 ft/min. Many of the above mentioned defects were detected. Moreover, it was determined that there is adequate uniqueness in the reflected signals from different defects that their type may also be distinguished. Microwave testing techniques are insensitive to many of the environmental limitations associated with manufacturing of rolled steel, and hence they are great candidates for this purpose. Clearly, the same techniques can be identically used when inspecting the surface of other rolled metals such as aluminum.

6.6 BARRIERS AND FUTURE

In the preceding sections and chapters in this text, many attributes of microwave non-destructive testing and evaluation techniques have been discussed. It is by now undoubtedly clear that these techniques are not only viable but they are uniquely applicable in many situations in which other non-destructive testing techniques do not apply. It is time to seriously consider utilizing these techniques in more practical applications than exploratory investigations.

Most individuals involved in non-destructive testing applications are still not very familiar with microwave techniques. Lack of clear understanding of the potential of these techniques, lack of available off-the-shelf microwave inspection systems, and some misunderstandings of the issues associated with these techniques (i.e. microwave power, signal interference, etc.) have contributed to this fact. As explained throughout this text, optimization is an invaluable feature of these techniques. However, a system with a set of measurement parameters (i.e. frequency of operation and standoff distance) usually is optimized for a specific environment. For the same system to be used in another application certain adjustments may be required. There are instances in which when microwave non-destructive testing techniques are deemed the most viable solution, the over expectation of its capabilities may become a detriment to its utility. Many industries, in need of robust inspection systems, are not always willing to invest in developing new microwave inspection systems. They rather acquire what they need off-the-shelf as they do other established non-destructive testing equipment. In some cases engineers involved in the day-to-day testing operations have recognized the unique potential of microwave techniques, however their managers who will have the final say may not have. This has also not helped to increase the utilization of these techniques. Bolomey outlined some of these issues (Bolomey, 1996).

For those who are involved in the research and development of microwave non-destructive testing and evaluation techniques the future is quite bright. There are several reasons for this optimism:

- The materials technology by which lighter and stronger composites are being manufactured to replace metals, will require robust testing and evaluation techniques for process and quality control. Microwave testing techniques are becoming more of an accepted candidate for these evaluation purposes, and will be more and more utilized in the future.
- The explosion in the wireless technology has been a great impetus for the increased availability of microwave and millimeter wave components. Such components and devices are much more readily and inexpensively available now than before. This will invariably result in more microwave inspection systems to be designed, developed and commercialized.
- Various governmental agencies have been investing research and development funds in microwave non-destructive technology. In some cases Small Business Innovative Research grants have been considered for microwave non-destructive testing and evaluation technologies. It is only a matter of time before some of these investments will bare fruits in the form of commercialized microwave inspection systems.
- It is becoming apparent to many interested individuals that obtaining a robust inspection may require the fusion of several distinct testing techniques. It is anticipated that in many of such applications microwave testing techniques will play a prominent role.

It is the author's hope that the materials presented in this text have been useful to many individuals. I also hope that this text has provided for a forum to better understand the value of this technology and has aided in increasing their practical utilization.

REFERENCES

Bakhtiari, S., Ganchev, S. and Zoughi, R. (1994) Analysis of the radiation of an open-ended coaxial line into stratified dielectrics. *IEEE Transactions on Microwave Theory and Techniques*, **42** (7), pp. 1261-67.

Benally, A.D., Bois, K., Nowak, P.S. *et al.*, (1999) Microwave nondestructive detection of chloride in cement based materials. *Proceedings of the Ninth International Symposium on Nondestructive Characterization of Materials*, Sydney, Australia.

Bois, K. (1999) Near-Field microwave inspection and characterization of cement based materials. Ph.D. Dissertation, Electrical and Computer Engineering Department, Colorado State University, January, Ft. Collins, CO.

Bois, K., Benally, A.D., Nowak, P.S. *et al.* (1999a) Application of near-field microwave sensing techniques for segregation detection in concrete members. *Proceedings of the Review of Quantitative Nondestructive Evaluation*, **19**.

Bois, K., Benally, A.D., Nowak, P.S. *et al.*, (1999b) Dielectric properties of concrete at S- and X-band: a near-field investigation. *Proceedings of SPIE, Subsurface Sensors and Applications Conference*, **3752**, Denver, CO.

Bois, K., Benally, A.D., Nowak, P.S. *et al.* (1998a) Cure-state monitoring and water-to-cement ratio determination of fresh Portland cement based materials using near-field microwave techniques. *IEEE Transactions on Instrumentation and Measurement*, **47** (3), pp. 628-37.

Bois, K., Benally, A.D. and Zoughi, R. (1998) Near-field microwave scattering analysis of concrete materials suing open-ended waveguide at S- and X-band. *Proceedings of the SPIE Symposium on Nondestructive Evaluation Techniques for Aging Infrastructure Manufacturing*, **3396**, pp. 37-46.

Bois, K., Benally, A.D., Nowak, P.S. *et al.*, (1997) Microwave nondestructive determination of sand-to-cement (s/c) ratio in mortar. *Research in Nondestructive Evaluation*, **9** (4), pp. 227-38.

Bois, K., Campbell, H., Benally, A.D. *et al.*, (1998b) Microwave noninvasive detection of grout in masonry. *Masonry Journal*, **16** (1), pp. 49-54.

Bois, K., Mirshahi, R. And Zoughi, R. (1997) Dielectric mixing models for cement based materials. *Proceedings of the Review of Quantitative Nondestructive Evaluation*, **16A**, pp. 657-63.

Bolomey, J.Ch. (1996) Some aspects related to the transfer of microwave sensing technology. *Proceedings of the Materials Research Society Symposium*, **430**, pp. 53-58.

Bolomey, J.Ch. and Joachimowicz, N. (1994) Dielectric metrology via microwave tomography: present and future. *Materials Research Society Proceedings*, **347**, pp. 259-68.

Bolomey, J.Ch. and Pichot, C. (1990) Microwave tomography: from theory to practical imaging systems. *International Journal of Imaging Systems and Technology*, **2**, pp. 144-56.

Ganchev, S., Qaddoumi, N., Bakhtiari, S. *et al.* (1995) Calibration and measurement of dielectric properties of finite thickness composite sheets with open-ended coaxial sensors. *IEEE Transactions on Instrumentation and Measurement*, **44** (6), pp. 1023-29.

Joisel, A., Bois, K., Benally, A.D. *et al.*, (1999) Embedded modulated dipole scattering for near-field microwave inspection of concrete:

preliminary results. *Proceedings of SPIE, Subsurface Sensors and Applications Conference*, **3752**, Denver, CO.

Shalaby, W. and Zoughi, R. (1995) Analysis of monopole sensors for cement paste compressive strength estimation. *Research in Nondestructive Evaluation*, **7** (2/3), pp. 101-05.

Wang, Y. (1999) Analysis of surface crack detection in metals using coaxial sensors. Ph.D. Dissertation, Department of Electrical and Computer Engineering, May, Ft. Collins, CO.

Zoughi, R., Cone, G.L. and Nowak, P.S. (1991) Microwave nondestructive detection of rebars in concrete slabs. *Materials Evaluation*, **49** (11), pp. 1385-88.

Zoughi, R., Gray, S. and Nowak, P.S. (1995) Microwave nondestructive estimation of cement paste compressive strength. *ACI Materials Journal*, **92** (1), pp. 64-70.

Zoughi, R., Hayes, K. And Ganchev, S. (1996) Microwave detection of hairline surface-breaking cracks in metals using open-ended coaxial sensors: preliminary results. *Proceedings SPIE Symposium, Nondestructive Evaluation Techniques for Aging Infrastructure and Manufacturing Conferences*, **2945**, pp. 444-50, Scottsdale, AZ.

Zoughi, R., Huber, C., Ganchev, S. *et al.* (1997a) Rolled steel surface inspection using microwave methods. *Proceedings of the Eighth International Symposium on Nondestructive Characterization of Materials*, **VIII**, pp. 285-90, Boulder, CO.

Zoughi, R., Huber, C., Qaddoumi, N. *et al.* (1997b) Real-time and on-line microwave inspection of surface defects in rolled steel. *Proceedings of the Asia-Pacific Microwave Conference, APCM'97*, pp. 1081-84, Hong Kong.

Zoughi, R. Nast, T. and Nowak, P.S. (1992) Preliminary results of microwave reflectometry as a nondestructive tool for studying concrete properties. *Proceedings of the Conference on Nondestructive Evaluation of Civil Structures and Materials*, pp. 339-43, Boulder, CO.

Zoughi, R., Nowak, P.S., Bois, K. *et al.* (1998) Near-field microwave inspection of cement based materials - Microwave sensor for nondestructive and non-contact estimation of concrete compressive strength. Final Report, NSF Contract no. CMS-9523264 and EPRI Contract no. WO 9031-09, p. 359.

Index

GPSR Compliance
The European Union's (EU) General Product Safety Regulation (GPSR) is a set of rules that requires consumer products to be safe and our obligations to ensure this.

If you have any concerns about our products, you can contact us on

ProductSafety@springernature.com

In case Publisher is established outside the EU, the EU authorized representative is:

Springer Nature Customer Service Center GmbH
Europaplatz 3
69115 Heidelberg, Germany

www.ingramcontent.com/pod-product-compliance
Ingram Content Group UK Ltd.
Pitfield, Milton Keynes, MK11 3LW, UK
UKHW020133160926
475607UK00003B/18

9780412625008